Reconfigurable Computing Systems Engineering

Systems Engineering

Virtualization of Computing Architecture

Reconfigurable Computing Systems Engineering

Systems Engineering

Virtualization of Computing Architecture

Lev Kirischian

CRC Press
Taylor & Francis Group
Boca Raton London New York

CRC Press is an imprint of the
Taylor & Francis Group, an **informa** business

CRC Press
Taylor & Francis Group
6000 Broken Sound Parkway NW, Suite 300
Boca Raton, FL 33487-2742

First issued in hardback 2019

© 2016 by Taylor & Francis Group, LLC
CRC Press is an imprint of Taylor & Francis Group, an Informa business

No claim to original U.S. Government works

ISBN-13: 978-1-4398-5621-5 (hbk)

Library of Congress Cataloging-in-Publication Data

Names: Kirischian, Lev, author.
Title: Reconfigurable computing systems engineering : virtualization of computing architecture / author, Lev Kirischian.
Description: Boca Raton : Taylor & Francis Group, 2016. | Includes bibliographical references and index.
Identifiers: LCCN 2015042956 | ISBN 9781439856215 (alk. paper)
Subjects: LCSH: Virtual computer systems. | Adaptive computing systems.
Classification: LCC QA76.9.V5 K57 2016 | DDC 005.4/3--dc23
LC record available at http://lccn.loc.gov/2015042956

Visit the Taylor & Francis Web site at
http://www.taylorandfrancis.com

and the CRC Press Web site at
http://www.crcpress.com

To my mother, Margarita

and to my beloved family

Irina, Valeri, and Nina

Contents

Foreword

I am delighted to write the foreword for this book, which I believe can be very useful for system-on-chip architects and hardware designers of field programmable gate array (FPGA)-based systems.

The focal point of this book is the concept of hardware architecture virtualization. This concept assumes the representation of hardware circuits in the form of configuration information that will be deployed in programmable logic devices only when needed. This concept provides a significant increase in cost efficiency for FPGA-based systems, especially embedded stream-processing computing systems. Likewise, the concept of memory virtualization can significantly reduce the cost of general-purpose computers as it allows for the affordable implementation of complex software systems in platforms employing virtual memory organization. A further increase in cost efficiency for computing platforms is expected if the concept of virtualization would be extended to all resources including data processing, control, communication, and interface resources.

Fortunately, the progress of deep-submicron complementary metal-oxide-semiconductor (CMOS) technologies in FPGA devices has allowed the implementation of complex multimodal system-on-programmable chips (SoPCs). Nowadays, modern FPGAs are difficult to call field programmable gate array devices because they represent a set of arrays for different types of resources-on-chip. Indeed, most FPGAs contain arrays for configurable logic units, communication resources, digital signal processing blocks, embedded memory units, and interface resources. Thus, the term "field of configurable resources" (FCR), used by the author, seems more suitable for the configurable part of any reconfigurable computing platform.

However, most existing methods for system-on-chip design, for static random access memory (SRAM)-based FPGAs, are still very similar to the methods used in the design of large application-specific integrated circuits (ASICs). In other words, all hardware circuits to be used in any mode of operation are deployed in the FPGA as one "monolithic" circuit to be configured in FPGAs at the time of start-up only. This method of design makes systems-on-chip more and more complex, following the amount of available resources in growing microarchitectures of larger FPGA platforms. The compilation time of SoPC designs targeting FPGAs with 20–30 million equivalent system gates may take half a day. In addition, the cost efficiency of SoPCs implementing multimodal applications may decrease significantly. This is due to the fact that large portions of the hardware resources may not be used at all times and are associated with nonactive modes of the application.

Understanding this problem, most FPGA vendors have included new features in their computer-aided design (CAD) systems related to partial run-time reconfiguration of the set of FPGA resources. These novel features, coupled with respective hardware support in partially reconfigurable FPGA devices, have enabled the utilization of new principles and methods for the design of the complex SoPC based on partial run-time reconfiguration of the FPGA. This "opens the door" to the idea of effectively implementing resource virtualization in dynamically and partially reconfigurable FPGA devices. Unfortunately, the design methods for partial SoPC deployment in FPGA devices are not known to most hardware designers.

In this light, the material in this book will help to better understand the conceptually important fundamentals needed for the development and design of this new class of computing systems. In turn, it can create a new "mentality" for SoPC architects and hardware design engineers.

The systematic representation of material, the detailed discussion of the main concepts, and the analysis of pros/cons of different approaches in architecture virtualization and static and dynamic integration will make this book very useful for engineers and students specializing in the development and design of FPGA-based reconfigurable computing systems.

Now is definitely the right time to provide a new vision into system architecture as dynamic architecture consisting of two parts: the part deployed in hardware, which is needed for operation, and the part stored in memory, which does not consume hardware and power resources. Unfortunately, this type of vision into system architecture is quite rare, in the mind of designers.

As I look forward, it is my hope that this book can mitigate this problem and provide a conceptual vision and understanding of the fundamental aspects of reconfigurable systems organization and design.

Dr. Karen Safaryan
Vice President and CTO, Unique Broadband Systems, Ltd.
Director R&D and Engineering, UBS-Axcera Inc., Vaughan, Ontario, Canada

Preface

It is the pervading law of all things organic and inorganic, of all things physical and metaphysical, of all things human and all things superhuman, of all true manifestations of the head, of the heart, of the soul, that the life is recognizable in its expression, that form ever follows function. This is the law.

—Louis H. Sullivan (March 1896)*

During the last decade, progress in deep-submicron CMOS technology elevated the volume of hardware resources deployed in programmable logic devices at least in order of magnitude. This quantitative growth caused a significant change in the vision of system architecture design. Indeed, previously the hardware circuits allocated in programmable logic devices (PLDs) were considered the "glue logic" between other system components or dedicated processing circuits for custom functions. However, in recent years it has become possible to deploy a complete system-on-chip (SoC) in a single field programmable gate array (FPGA) device. In turn, this fact demanded the transformation of hardware design engineers into system architects for effective development of these SoCs. However, the existing university curriculum for computer systems engineers in most cases does not cover all aspects of system architecture synthesis, design, and implementation.

Motivation for Writing This Book

Novel features in CAD systems supported by the microarchitecture of modern FPGAs enabled the design of dynamic partially reconfigurable systems-on-programmable chips (SoPCs). These features have significantly changed viewing computing architecture as a static entity. The vision of the SoPC architecture came closer to the concept of "space–time continuum" where time multiplexing of hardware resources allows flexible transformation of the area of used resources to the duration of data execution and vice versa. As a result, the run time adaptivity of the computing system to variations in workload and environment became possible.

---

* Reprinted from L. H. Sullivan, The tall office building artistically considered, *Lippincott's Magazine*, 57, 403–409, March 1896.

The understanding of the efficiency of dynamic *architecture-to-task adaptation* in contrast to traditional *task adaptation* (programming) for a computing platform with static architecture has generated extensive research in the area of *reconfigurable computing systems* (RCS) since the early 1990s. However, only recently the concept of run time partially reconfigurable computing systems has founded its adepts in the industry associated with high-performance embedded computing systems. Following growing interest in this concept, major vendors of FPGA devices have started the implementation of special features to their CAD systems for FPGA devices supporting dynamic partial reconfiguration of on-chip resources. These new features have enabled creation of computing systems for which architecture is not entirely present in hardware circuits for the entire time. Instead, this type of architecture consists of two parts: (1) the part that is temporarily deployed in the hardware platform (e.g., FPGA) and (2) the part that is stored in different levels of a configuration memory hierarchy in the form of configuration bit files (bit streams). In other words, a large portion of computing architecture exists in a *virtual form* of files to be loaded in configurable logic devices only when needed and for the period while needed. This approach allows a significant reduction of system cost and power consumption because architectural components stored in memory as information objects do not occupy expensive hardware resources and consume very low power (if any) when not in use. This approach is very similar to the concept of *virtual memory*, where a task is not fully loaded to the electronic memory attached to the central processing unit (CPU). Instead, only those segments of the task that are currently requested for execution or expected to be requested soon are loaded to the electronic memory. The rest of the task segments are stored in different levels of memory hierarchy according to the probability of their initiation for data execution. This allows keeping most of the task segments (those not in use) in less expensive (per bit) secondary storages (e.g., hard-disk drives or solid-state drives) instead of expensive SRAM-based cache memory.

Unfortunately, the concept of architecture virtualization and associated methods for RCS synthesis and design are not well known to designers of FPGA-based systems. Most of them still consider SRAM-based FPGA devices as a programmable replacement of an application-specific integrated circuit (ASIC). Respectively, the design of SoCs is based on methods similar to the ASIC design where the entire architecture is deployed in one large circuit to be programmed into the FPGA at start-up time.

Therefore, the primary motive to write this book was to provide hardware design engineers, system architects, and students specializing in designing FPGA-based embedded systems with material to expand their vision of novel concepts in architecture organization and architecture virtualization in reconfigurable computing systems.

Organization and Readership for This Book

It was clear from the beginning that it would be very difficult or even impossible to consider all possible ways and methods of architecture organization of statically and dynamically reconfigurable computing systems. Thus, the emphasis of the book is describing the fundamental aspects of architecture organization in RCS (existing and potential) and analysis of pros and cons of different ways of RCS architecture implementation in on-chip and system levels. These fundamental aspects are described generally considering programmable logic devices (PLD) in their generalized organization. It is assumed that specifics of microarchitecture organization of FPGA devices or complex programmable logic devices (CPLDs) can change in a relatively short time. Indeed, it was possible to see significant changes in the microarchitecture of FPGA families developed by a same vendor (e.g., Xilinx Inc.) in the late 1990s in comparison to column-based microarchitecture of FPGAs developed in the early 2000s and then tile-based micro-architecture of FPGAs developed in 2010s. There is no doubt that this trend will continue in the future. On the other hand, fundamental principles and concepts of RCS architecture organization are much more stable over time. Therefore, the description of these concepts and principles in the book in a general form will give readers better understanding of the fundamentals of RCS organization not clouded by specifics of current organization of FPGA microarchitecture.

Nonetheless, keeping in mind the potential practical use of the book, all examples and literature sources are based on current engineering practice and existing programmable logic devices. These literature sources include original patents, official datasheets, user guides, and application notes provided by leading vendors of FPGA devices and CPLDs (Xilinx Inc. and Altera Corp.). In addition, the illustrative examples also use device-specific information retrieved from datasheets of current families of FPGA devices and CPLDs. The above approach allows demonstration of use the considered general principles and concepts on the current practically used configurable devices.

The material presented in the book is based on lecture notes of a graduate course, "Reconfigurable Computing Systems Engineering," which has been developed taught by the author at Ryerson University, Toronto, Canada, since 2008. The feedback from graduate students was used for determining the book content and the form of material presentation.

The book consists of 10 chapters, and the course material was planned to be taught within one semester (12 weeks) as a graduate course.

The book is oriented to students and engineers who have graduated with a bachelor's degree in computer engineering. No additional courses are a prerequisite. However, the book can also be used as a reference book by system architects working in the area of embedded reconfigurable systems design.

Structure of Each Chapter

Each chapter has the same framework, starting with an introduction and ending with a summary of the chapter content. The introduction briefly observes the main aspects presented in the chapter. Then, these aspects are discussed in detail in subsequent sections according to logical progression. The main terms defined in each chapter are used in other chapters in this book, but may not be those commonly used in other areas of technology. The definitions of terms, where possible, are made in layman's language for easier understanding. This approach of presenting content has shown higher efficiency in comparison to presenting formal definitions. Important words in sentences are emphasized by italic or bold fonts.

The presentation of each concept starts from declaration of the idea on which the concept is based followed by a general description of this concept, which is illustrated by a quantitative example. Each chapter ends with questions, exercises, and problems that allow self-testing of all parts of the material presented in the chapter. The sequence of questions and problems follow the sequence of material presentation in the chapter. Thus, all answers can be found directly in a chapter's text.

The literature referred to is listed at the end of each chapter and is recommended for a more detailed consideration of material presented in a chapter. Due to space constraints, FPGA device specifics and implementation methods of principles in actual FPGA devices or CPLDs are referred to respective official datasheets, user guides, application notes, etc. It is strongly recommended these sources from the literature be read in addition to the material provided in each chapter. This practice will allow readers to acquire a "binocular" vision for each concept, including physical implementation details.

Acknowledgments

First, the book would have been difficult to develop without the efforts of all members of the Embedded Reconfigurable Systems Laboratory (ERSL) at Ryerson University, Toronto, Canada. Most of R&D projects conducted at ERSL for almost two decades directly influenced the book contents. These projects have expanded our conceptual understanding of RCS architecture organization for different industrial applications, associated design experience, and, therefore, the vision of modern RCS architecture organization.

These projects would never have been successful without funding support provided by several federal and provincial agencies: Natural Sciences and Engineering Research Council of Canada (NSERC), Ontario Centres of Excellence (OCE), FedDev Ontario, Materials and Manufacturing Ontario (MMO), Centre of Information and Communication Technology (CCIT), and our industrial partners MDA Space Missions (former SPAR Aerospace), Unique Broadband Systems (UBS Ltd.), and Soft-R US Inc.

In addition, it was very important to have equipment and software support in the form of high-performance workstations, modern instrumental software and CAD tools, FPGA and CPLD evaluation boards, FPGA devices, and device programmers and test equipment for experimental setups provided by Canadian Microelectronic Corporation (CMC Microsystems), Altera Corp., Xilinx Inc., and UBS Ltd.

Special mention should be given to the Faculty of Engineering and Architectural Science (FEAS) at Ryerson University for providing lab facilities, start-up funding for ERSL, and continuous high-quality technical support.

Personally, I would like to thank Dr. Karen Safaryan, vice-president and CTO at UBS Ltd. and director of R&D and Engineering at UBS-Axcera Inc. for his support of our research and development projects for many years. He was always ready to share his knowledge, expertise, and vision with all researchers and engineers at ERSL. Regardless of his very busy schedule, he was able to find time for our long discussions regarding the necessity and contents of the graduate course and this book. These discussions became the ignition point for developing the graduate course and then the book. I also would like to thank Dr. Karen Safaryan for his foreword for this book.

I would like to express my sincere gratitude to Dr. Victor Dumitriu and Dr. Valeri Kirischian for reviewing the book chapters and for their valuable comments and corrections in the text. Their assistance allowed me to have a "second opinion" that is so important in any development project, including the development of the textbook.

Also, I would like to thank all graduate students who provided me with their feedback, suggestions, and recommendations regarding the course contents and presentation details. All their comments and suggestions have been considered in detail and are reflected in the book.

The final and greatest thanks I would like to address to my family for the patience, selfless support in all the meanings of this word, and hope that all my efforts in "endless" writing, re-writing, and correcting chapters eventually will be useful for the readers of this book.

Therefore, I am dedicating this book to my family!

Lev G. Kirischian

Author

Lev G. Kirischian received his BASc and MASc degrees in electrical engineering from the Moscow Institute of Aviation Technology (MAI), Moscow, Russia, in 1978 with specialization in aerospace control systems. He obtained his PhD degree in computer engineering from the Moscow Power Engineering Institute (MPEI) in 1985 with specialization in the area of parallel and reconfigurable computing systems.

Since 1978, Lev Kirischian has been involved in R&D projects on first generation of Soviet parallel computers with reconfigurable architectures: PS-300 (1979), PS-2000 (1982), and PS-3000 (1984). His PhD research was focused on automated dynamic selection of architecture configurations for computing systems with reconfigurable structures. This work was conducted during his PhD study (1981–1985) in cooperation with the Institute of Control Sciences (Moscow), the leading research organization for R&D of high-performance parallel computing systems in the USSR.

During 1978–1986, Lev Kirischian was the project leader and developer of digital signal acquisition modules for the first distributed microprocessor-based control system, M-64, for the new generation of Soviet nuclear power plants based on nuclear reactors VVER-1000 (Water–Water Energetic Reactor—1000 MW).

During 1986–1995, Dr. Kirischian was a professor at the Institute of Professional Development for Industrial Electrical and Computer Engineers, Tbilisi, Georgia.

Since 1995, Lev Kirischian has been working in Canada. He was a member of the R&D group at Unique Systems Inc., which developed the first digital audio broadcasting (DAB) system in North America. The core of this system was the FPGA-based coded orthogonal frequency-division multiplexing (COFDM) modulator. Later, this type of modulator was used in Sirius-XM satellite radio broadcasting transmitters as well as in different digital video broadcasting (DVB) systems (DVB-T and DVB-SH) manufactured and deployed around the globe by Unique Broadband Systems (UBS) Ltd.

Dr. Kirischian joined Ryerson University (Toronto) in 1998. His research interests are in the area of dynamically reconfigurable computing systems, automated architectural synthesis of data-stream computers, and workload adaptive and self-healing reconfigurable architectures. In the last decade, he has developed and taught several courses (both undergraduate and graduate levels) associated with high-performance and reconfigurable computing as well as high-level synthesis of application-specific processors.

In 1999, Dr. Kirischian established the Embedded Reconfigurable System Laboratory (ERSL) at Ryerson University to conduct R&D projects for research and industrial purposes. All projects conducted at the ERSL are focused on the research and development of run-time reconfigurable computing systems for the next generation of self-restorable space-borne embedded computer platforms, run-time adaptive DVB systems for satellite and terrestrial networks (DVB-T, DVB-SH, etc.), and 3D-panoramic machine vision systems. These projects have been undertaken in cooperation with several academic and industrial R&D groups at the University of Toronto (Toronto), Queen's University (Kingston), MDA Space Missions (Brampton), Unique Broadband Systems Ltd. (Concord), and Canadian Microelectronic Corporation (Ottawa).

Dr. Lev Kirischian is a registered professional engineer (P. Eng.) and a member of the IEEE. He has served as a reviewer for NSERC discovery grants, NSERC collaborative research and development (CRD), and NSERC strategic partnership grant programs as well as for several international journals, conferences, and workshops.

List of Abbreviations

ADC	Analog-to-digital conversion
ALU	Arithmetic and logic unit
AMBA	Advanced microcontroller bus architecture
ARM	Advanced RISC machine
ASI	Asynchronous serial interface
ASIC	Application-specific integrated circuit
ASIP	Application-specific instruction processor
ASP	Application-specific processor
ASVP	Application-specific virtual processor
AU	Arithmetic unit
AXI	Advanced eXtendable interface
BCR	Base configurable region
BIST	Built-in self-test
BLE	Basic logic element
BW	Bandwidth
CAD	Computer-aided design
CAN	Controller area network
CCU	Configurable communication unit
CDB	Check-data buffer
CF	Configuration frame
CGRA	Coarse grained reconfigurable array
CISC	Complex instruction set computer
CLB	Configurable logic block
CMOS	Complementary metal–oxide–semiconductor
CoDec	Compression/decompression
COFDM	Coded orthogonal frequency-division multiplexing
CPE	Configurable processing element
CPLD	Complex programmable logic device
CPU	Central processing unit
CRC	Cyclic redundancy check
CS	Chip select
CSPP	Computing system with programmable procedure
DAC	Digital-to-analog conversion
DCI	Digital controlled impedance
DDR	Double data rate
DFG	Data flow graph
DLL	Delay-locked loop
DLP	Data-level parallelism
DMA	Direct memory access

DPR	Dynamic partial reconfiguration
DRCS	Dynamically reconfigurable computing system
DSP	Digital signal processing
EDVAC	Electronic discrete variable automatic computer
FAR	Frame address register
FCR	Field of configurable resources
FFT	Fast Fourier transform
FIR	Finite impulse response
FP	Floating point
FPGA	Field programmable gate array
FPU	Function processing unit
FSM	Finite state machine
GPIO	General purpose input/output
HDL	Hardware description language
HSTL	High-speed transceiver logic
I/O	Input/output
I2C	Inter-integrated circuit
IBUF	Input buffer
ICAP	Internal configuration access port
ICL	Internal controller-loader
IIR	Infinite impulse response
IOE	Input/output element
IP-TV	Internet protocol television
JPEG	Joint Photographic Experts Group
LED	Light-emitting diode
LUT	Look-up table
LVCMOS	Low-voltage CMOS
LVDS	Low-voltage different signaling
LVTTL	Low-voltage transistor–transistor logic
MAC	Multiply-accumulate
ME	Memory element
MIMD	Multiple instruction, multiple data
MISD	Multiple instruction, single data
MOSFET	Metal–oxide–semiconductor field effect transistor
MPEG	Motion Picture Expert Group
MUX	Multiplexor
NOP	No operation
OBUF	Output buffer
OBUFT	Output buffer with tristate circuit
Op-Amp	Operational amplifier
OTS	Off-the-shelf
PAR	Place and route
PCB	Printed circuit board
PCI	Peripheral component interface

PE	Processing element
PLA	Programmable logic array
PLB	Programmable logic block
PLD	Programmable logic device
PLL	Phase-locked loop
PPE	Programmable processing element
PR	Partially reconfigurable
PRR	Partially reconfigurable region
PSE	Programmable switching element
PSM	Programmable switch matrix
PWM	Pulse-width modulation
R&D	Research and Development
RAM	Random access memory
RASP	Reconfigurable application-specific processor
RCC	Resource configuration controller
RCS	Reconfigurable computing system
RF	Radio frequency
RISC	Reduced instruction set computer
RMS	Reconfiguration management system
RP	Reconfigurable partition
RPU	Reconfigurable processing unit
RTR	Run-time reconfigurable
RUF	Resource utilization factor
SDRAM	Synchronous dynamic random access memory
SerDes	Serializer/deserializer
SFCR	Statically configurable field of configurable resources
SFMD	Single function, multiple data
SG	Sequencing graph
SI	Signal integrity
SIMD	Single instruction, multiple data
SISD	Single instruction, single data
SoC	System-on-chip
SoPC	System-on-programmable chip
SPDT	Single pole, double throw
SPI	Serial peripheral interface
SRAM	Static random access memory
SSTL	Stub series terminated logic
TCP-IP	Transmission control protocol/internet protocol
TMR	Triple modular redundancy
UART	Universal asynchronous receiver/transmitter
USB	Universal serial bus
VC	Virtual component
VHbC	Virtual hybrid component
VHC	Virtual hardware component

VHDL	Very-high-speed logic hardware description language
VLSI	Very-large-scale integration
VM	Virtual memory
VSC	Virtual software component
XGA	eXtended graphics array

1

Introduction to Reconfigurable Computing Systems

1.1 Introduction

System architecture reconfiguration is a quite usual way of adapting to environmental changes in many technical systems and in life forms of evolution. It could be seen in naval systems (e.g., reconfiguration of sails to adapt to the wind direction or weather conditions) and aviation (e.g., aircrafts with reconfigurable wings or wing's elements). Nowadays, the same approach is used in flexible (reconfigurable) manufacturing systems, robots, and many other areas. The same way of adaptation to environmental changes can be found in almost all biological systems and their evolution on Earth over billions of years. Therefore, it is difficult to imagine that computing systems, being flexible in nature, could avoid the aforementioned concept of architecture reconfiguration in response to variations in the workload or external conditions.

Initially, computing system architecture was considered as fixed from the hardware point of view (e.g., classical von Neumann computing architecture [1]). However, tasks (applications) were obviously different. Therefore, *adaptation* of the variable task to a fixed computing architecture was necessary. This adaptation process was called *task programming* and had to be done by a programmer. In other words, task programming and program compilation processes were *task-to-architecture* adaptation processes.

The concept of reconfigurable computing machines assumed the opposite process of *architecture-to-task* adaptation. This concept was proposed by Gerald Estrin in 1960 [2] in the form of a computing machine with fixed and variable architectural components. A more detailed paradigm was presented by Estrin, Bussell, Turn, and Bibb in 1963 as a *restructurable* computer system for parallel processing [3]. Later, R. Hartenstein called this paradigm an "antimachine" [4]; its meaning is opposite to von Neumann's paradigm of task-to-architecture adaptation.

It is necessary to mention that the usual association of reconfigurable computing systems (RCSs) with field-programmable gate array (FPGA) devices is not quite correct. First, the history of RCS had started at least 20 years prior to the concept of a look-up-table (LUT)-based programmable logic device (PLD) being proposed and implemented in the form of an FPGA. The RCS paradigm was initiated by a conceptual understanding of the direct correspondence between computing architecture and task algorithm and data structure. On the other hand, FPGA devices became an efficient hardware platform for different types of RCS due to their *homogenous organization* of configurable logic cells and communication infrastructure.

The evolution of RCS within the last two decades (from the early 1990s) has shown a very high cost-efficiency and power efficiency, as well as performance acceleration of 2–3 orders of magnitude compared to conventional complex instruction set computing (CISC) and even reduced instruction set computer (RISC) systems [5]. Nevertheless, RCSs are still quite exotic in the world of computing machines. There are several reasons for that. And the major reason is the nature of the RCS development process associated with utilization of the concept of architecture-to-task adaptation instead of the usual task-to-architecture programming.

On the other hand, the physical constraints of sequential execution of information in synchronous digital circuits pushed computing system developers to the parallel processing of data-streams. The concept of parallel data-processing also is not new. The idea of *multiprocessing* was in development from the early 1960s. In 1966, Michael J. Flynn proposed the taxonomy for computing machines based on the number of streams of data and instructions [6]. However, it was found that the effectiveness of a fixed architecture of a parallel computing machine strongly depends on the classes of applications and the ability to utilize their natural parallelism in algorithm(s) and data structure(s). At the same time, the necessity to accelerate task execution with variable (multimodal) algorithms and/or data structure caused wider utilization of the RCS concepts. Furthermore, progress in the submicron technology of electronic circuits provided efficient computing platforms with run-time reprogrammable logic circuits with very high capacity of on-chip computing, communication, and interfacing resources. The aforementioned trends made RCS a cost-effective solution for a wide variety of industrial and scientific applications from video-processing and broadcasting applications to digital signal processing (DSP), flexible automation, robotics, data mining, etc.

Nowadays, after almost three decades of RCS evolution, the area of RCSs and their applications is very wide and thus cannot be fully described in one book. However, there is an aspect that could be considered as one of the most important things in understanding the concept of RCS and the methodology of architecture development of this class of computing machines. This is the aspect of *virtualization of computing architecture*.

The concept of hardware virtualization is based on the same idea as in any hardware description language, where logic circuits can be completely described by a limited set of operators. In other words, any data-processing, data communication, and interface circuit can be represented in the form of a data-file. This could be put in analogy with *chromosomes*, where a structurally organized set of DNAs *programs* the physical organization and functions of living organisms. The aforementioned concept of architecture virtualization, being implemented on the platform of PLDs, allows a significant increase in the cost-efficiency of computing machines because only components that are actually necessary for data-execution, data storage, and data-transfer are present in the form of electronic circuits in the hardware platform. All other components could be stored in special memory that keeps configuration bit-files of all possible components.

The concept of architecture virtualization can be considered as an extension of the concept of memory virtualization in modern computers. The concept of virtual memory (VM) was proposed (to the best of our knowledge) in 1956 and assumed placement to the most expensive memory (random access memory) only those segments of the task that were expected for execution in the near future [7]. Thus, the memory resources, being separated by their volume-to-cost ratio and used in respect to the *close-to-execution* order, allowed a significant increase in the cost-efficiency of the entire computing system. This concept was possible to implement due to homogenous nature of memory units and standardized communication infrastructure (buses and protocols) deployed in modern computing systems.

Further unification of basic computing resources in the form of configurable logic blocks in the FPGAs and complex programmable logic devices (CPLDs) made possible the utilization of principles of *hardware virtualization for computing, communication,* and *interface resources* on the base of these reconfigurable hardware platforms. Nonetheless, the development of computing systems on the base of FPGAs and CPLDs still is based on the principles and methodologies of the design of application-specific integrated circuits (ASICs), where all functionalities of the system are directly implemented in the actual hardware and configured at start-up time.

Therefore, to accelerate the utilization of principles of architecture virtualization for the development of high-performance and cost-efficient RCS, the course "Reconfigurable Computing Systems Engineering" was established and taught since 2008. Further, the book has been written and proposed for publication on the basis of this course.

Thus, the goal of this book is to provide readers with a systematic path to architecture development of RCS oriented toward computationally intensive multimodal applications (e.g., DSP, video/audio/image processing, data mining, multimedia applications).

The focus is placed on architecture organization and development of embedded RCS from the system-on-chip (SoC) level to the onboard and system-level organization of the RCS.

The book describes the basic principles of the organization of PLDs as the platform for RCS. In addition, the theoretical aspects of synthesis of the system architecture in spatial and temporal domains in the field of configurable resources (FCR) are discussed in details and with appropriate examples. The analytical representation of the material was minimized accordingly.

It was taken into consideration that the microarchitecture of FPGA devices and CPLDs may change in time. Therefore, the material is presented with general architecture of PLDs in mind. However, specific examples based on recent FPGA architecture organizations are considered for illustration and better understanding of the material.

Special focus is placed on hardware *virtualization from component level to architecture integration*. The architecture integration process is considered in static and dynamic forms for statically and dynamically reconfigurable RCS architectures.

The book, however, does not consider many other aspects associated with RCS development. The volume of the book and the limited time did not allow discussing the aspects associated with the software development process for the RCS as well as the organization of operating systems for RCS. As mentioned earlier, it would be very difficult or almost impossible to observe all aspects associated with RCS architecture organization and development. Nevertheless, the author believes that this book could give better understanding of the concepts and principles of architecture organization and development of the RCS.

1.2 Computational Process and Classification of Computing Architectures

Organization of the computing process is the key aspect that influences the organization of computer architecture. Thus, it is necessary to describe the data computing process in general terms. It is also necessary to define major terms associated with that process for further utilization in the text of this book as the meaning of terms may vary based on the area of application. For example, the term "page" in a textbook and in virtual memory (VM) has different meanings. It is also important to give a formal definition of computing architecture because this definition will then be used in further chapters of this book.

The data computing process in general consists of two flows:

1. *Flow of data to be computed*: The dataflow is usually considered to consist of a sequence of data-elements—operands, vectors of operands, etc.

2. *Flow of control information*: The flow of control information is considered to consist of a certain sequence of operations to be performed on associated data-elements in dataflow.

Therefore, the *task* can be defined as an *information object* that consists of the following:

1. *Algorithm* that represents the flow of control information to perform requested operations
2. *Data structure* that represents organization of data-elements associated with this algorithm

For example, if the task is represented by the following formula

$$Y = (A_i + B_i) \times K \quad \text{where } i = 1, 2, \ldots, 1024,$$

this means that there is a sequence of operations: addition, multiplication, pointer (i) increment, comparison (to 1024), and branch (jump). This sequence has to be executed in a certain order repeatedly, while all operands are organized in the form of A and B vectors.

Hence, digital computing system should consist of two major parts:

1. Data-processing path (data-path)
2. Control unit to provide data-path circuits with control information

In general, it is possible to have single or multiple flows of data and/or control information. Therefore, all possible computer architectures can be divided into four major classes according to the number of dataflows (or data-streams) and control information flows (streams).

The following classification was proposed by Michael J. Flynn [6] and is known as Flynn's taxonomy:

1. *Single instruction and single data* (SISD) stream computer architecture organization where control and data information are processed sequentially. This means that at any given moment, only one element of control information represented in the form of *instruction* is processed in the control unit and only one stream of data is executed in the data-path. Most of the microprocessors with CISC architectures belong to this class of computing machines. These types of computers are the most simple and low cost. However, the drawback of the SISD architecture is its relatively low performance compared with other classes of computing systems.
2. *Single instruction and multiple data* (SIMD) streams computer architecture organization where one stream of control information can be used for multiple parallel data-paths processing different streams

of data using the same algorithm. Utilization of this architecture requires natural independence of operands (data-elements) from each other to allow simultaneous execution. The cost–performance characteristic of these architectures is usually higher than SISD due to the minimization of hardware costs (only one control unit for many data-paths). At the same time, the performance of SIMD architectures can be increased according to the number of data-paths working in parallel. SIMD architectures are widely used in so-called array and vector computers.

3. *Multiple instruction and single data* (MISD) stream computer architecture organization where one data-stream is processed in parallel by different algorithms. There are limited classes of applications associated with real-time stream execution that may use MISD architectures. These applications may be associated with decryption of encrypted data-streams, parallel processing of data-streams received from the same sensor or set of sensors for different actuators, etc.

4. *Multiple instructions and multiple data* (MIMD) streams computer architecture organization where multiple processing elements (PEs) or complete processors with their own control units and memory modules are executing simultaneously different tasks or segments of the same task. This is the widest class of parallel computing system, ranging from multi-/many-core processors to distributed computing networks.

In summary, the SISD architecture can be considered as the only general-purpose computing architecture due to the sequential nature of data and control information processing. The architecture and information execution process in the SISD class of computers is based on the concept of a *universal Turing machine* proposed by Allan Turing in 1936 and published in 1937 in the classical paper titled "On Computable Numbers, with an Application to the Entscheidungsproblem" [8]. A few years later, the concept of SISD was implemented in the first *programmable electromechanical computer*, Z3, developed by Konrad Zuse in 1941 [9] and in the architecture of the first *programmable electronic computer*, electronic discrete variable automatic computer (EDVAC), proposed by John von Neumann in 1945 [1]. This architecture, called von Neumann's architecture, has become the de facto standard for most computers for at least two decades.

The remaining classes of computer architecture utilize different kinds of parallelism in processing data and/or control information and can, therefore, provide higher performance. However, each of the aforementioned classes of architectures is oriented toward a certain class of applications. It is also necessary to mention that these parallel architectures are more complex in development and design (compared with SISD machines) but can provide higher performance characteristics.

1.3 Formal Definition of Computing Architecture

As with any complex system, a computer consists of parts, each of which has its own functionality. These parts are usually called *components*. Traditionally, computer components were considered as *hardware* and *software* components. Thus, components can be represented (1) as a circuit in hardware components or (2) as control code (sequence of instructions) in software components.

As parts of a system, components need to exchange data and/or control information as well as feedback and synchronization signals. The information exchange process is organized via certain *ports* or *interfaces*. This set of ports or interfaces is also considered a part of the component.

Definition 1.1 *Component* is as an *information object* that can perform a determined *set of functions* and exchange information via determined *interface(s)*.

Nevertheless, the physical nature of a component represented in any form should always be implemented on the appropriate hardware platform: (1) hardware components in associated integrated circuits and (2) software components in the appropriate processor platform.

In addition to the aforementioned implementations, components can be implemented in the form of soft-core or *configurable* components to be deployed in the PLD as a platform. These soft-core components can be configured inside PLDs (e.g., FPGAs or CPLDs) by loading associated *configuration bit-files* to the configuration memory of PLD.

Transfer of information between components is organized via *links*. These can take the form of peer-to-peer (or point-to-point) links, multipoint, and point-to-multipoint links. In all cases, links perform functionality associated with predetermined *communication protocols*. In general, a communication protocol is considered to consist of a sequence of actions in the temporal domain associated with (1) establishing communication, (2) information exchange, and (3) termination of communication. In other words, links can also be considered as components with functionality specialized for information transfer. However, due to their specific role in architecture, links are usually considered as separate elements of the computing architecture.

Both components and links are implementing system functionalities in the spatial domain (in space). However, in both cases, the sequential processes can be considered as *procedures* in the temporal domain (in time). Thus, processes of data acquisition, data-transfer, data-processing, data storage, etc., are procedural elements of system architecture—*functional procedures*.

Procedures can be implemented directly in the form of hardware (e.g., finite state machines) or in the form of software (control code). Therefore, procedures are also *information objects* to be deployed on a specific hardware platform.

Therefore, it is possible to give the formal definition of the computing architecture as follows:

Definition 1.2 *Computing architecture* (A_{cs}) can be defined as a triple set of

1. Functional components (C_i)
2. Communication links between components (L_{ij})
3. Functional procedures associated with components and links (P_k)

$A_{cs} = \{C_i,\ L_{ij},\ P_k\}$, where $i, j = 1, 2, \ldots, N$; $k = 1, 2, \ldots, M$; here N represents the number of components and M represents the number of procedures in the system.

1.4 Correspondence between the Task and the Computing Architecture

As per the definition of a task given in Section 1.2, a task consists of an algorithm and a data structure. Obviously, a computing system should provide all operations required for the task algorithm. In addition, the data storage (memory) organization should reflect the data structure and specifics of data-elements (e.g., 32-bit vector registers for the array of 32-bit words). In other words, computer architecture must satisfy the specifics of the algorithm and data structure associated with a particular task. For instance, if there are two tasks represented by two formulas

1. Task 1: Matrix multiplication: $AB = [A] \times [B]$, where $[A]$ and $[B]$ are 4×4 matrices.
2. Task 2: $S = \sum_{i=0}^{n}(a-b)i/K$, where $i = 0, 1, \ldots, 255$.

A computer should provide "add" and "multiply" operations for the first task and "add," "subtract," and "divide" operations for the second. Thus, if the system can provide only "add" and "multiply" operations, it would not be able to execute task 2 due to the lack of a "divide" operation. Since different tasks may require different arithmetic, algebraic, logic, and other operations, computing architecture should have a lot of functional components to accommodate the aforementioned operational needs. On the other hand, due to the procedural nature of a task (data and control dependencies in the task algorithm), only one or a few of the aforementioned operations can be active at a time. In turn, this will increase system complexity and the associated cost of computer hardware reducing the cost-efficiency of the computing system. Therefore, it is necessary to have some cost-efficient mechanism to accommodate a computing architecture to the task. There can be three possible ways for conforming task and computing architecture:

1. Adaptation of variable tasks to a fixed computing architecture
2. Development of specific computing architectures for every (fixed) task
3. Adaptation of a variable computing architecture to a given task

---

1.5 Concept of a Computing System with a Programmable Procedure

Idea: If it would be possible to find a universal set of operations that permits the implementation of any operation as a procedure, utilization of hardware resources can be maximized and the hardware part of the architecture can be fixed.

This universal set of operations was named the set of *elementary (or primitive)* operations. Using such universal set of elementary operations, the concept of a universal *arithmetic and logic unit (ALU)* has been proposed. In other words, any data acquisition, data storage, arithmetic, logic, algebraic, or other operations can be represented as a sequence of the aforementioned elementary operations. In the case of the simplest ALU, it was determined that "arithmetic addition," "data shift" (right and left), logic "inversion," and logic "AND" operations can be considered as primitive.

If any operation in any task could be executed using a certain sequence of elementary operations, then the hardware part of the computing architecture (components and links) can be kept fixed for all tasks. In such a case, conformation of the task to the computing system can be done by composition of a specific procedure, which includes elementary operations for data acquisition, data-execution and data storage functions. This procedure is called a *program*. Loading the program to the program memory of the computer is called *programming*. Thus, programming became the mechanism for adaptation of the task algorithm and data structure to the computer architecture. Therefore, this class of computers can be defined as *computing systems with programmable procedure (CSPP)* or programmable computers:

$$A_{cspp} = \{C_i, L_{ij}, P_{\sim}\}, \quad \text{where } i, j = 1, 2, \ldots, N: P\text{-variable}$$

Fixed   Variable

There are several important advantages to the CSPP concept:

1. Relatively low cost for the hardware part of these systems. There are two reasons for that: (a) minimization of hardware cost due to a limited number of elementary operations implemented in the ALU and

(b) minimization of hardware cost by mass production of the same hardware platform for almost any application (task).

2. Flexibility of the computing platform due to the ability for rapid reprogramming of system functionality by loading a new procedure (program).

3. Relatively easy and low-cost composition process for data-processing procedures using *procedural programming languages*.

This list of benefits can definitely be continued. This is not necessary, however, as the three aforementioned advantages have already made computers with programmable procedures the most popular computing system type in both industry and the consumer market.

However, these types of computing systems have certain drawbacks, mostly associated with limited performance and reliability. Ironically, the reasons for these disadvantages look like a direct extension of the aforementioned benefits. The major limitations of the CSPP concept are based on (1) technological reasons, (2) reasons associated with control information processing overhead, and (3) reasons associated with the organization of procedural programming.

Technological reasons for performance constraints are as follows: (1) limited switching frequency of transistors that a certain technology can provide and (2) limited speed of signal propagation in a semiconductor. The switching frequency of transistors directly depends on manufacturing technology. If the technology provides smaller linear dimensions for transistors, the internal capacitance of these transistors is reduced and, therefore, allows higher switching frequency. However, the linear dimensions of transistors are restricted by both physical reasons and limitations in the manufacturing process. Therefore, switching frequency and the associated clock frequency have their physical limit. At the same time, the signal propagation speed also depends on many factors (e.g., type of semiconductor, impedance of routing). In any case, this speed can never reach the speed of light. In addition to that, the available power (consumption and dissipation) provided by a given technology is also a great limiting factor for performance acceleration.

Reasons associated with *control information processing* are also important limiting factors for increasing the CSPP performance. As mentioned earlier, the CSPP concept is based on the utilization of elementary operations as the basis of the computing platform. In other words, the entire task is divided into a sequence of elementary subtasks, called *instructions*. The instruction, as an *elementary task*, consists of one *elementary operation* and *elementary data structure*—a pair of data-elements (*operands*) and one result. Therefore, execution of the instruction requires processing of several steps associated with initiation, source-data delivery, data-processing, and result storage. For most CSPPs, these are the stages of instruction processing [10]:

1. *Instruction fetch* (IF)—fetch the instruction word from the memory
2. *Instruction decode* (ID)—decode the instruction word to determine the operation code, location of operands, and location of result
3. *Data fetch* (DF)—fetch operands and store them in ALU registers
4. *Execution of operation* (EXE)—execute the operation
5. *Store result* (SR)—store the result in registers or memory

Thus, one operation on data-elements requires several operations on control information encoded in the instruction word. It means that in addition to ALU (data-processing unit), the CSPP should have a separate processor for control information. This control information processor was called the *control unit*. The combination of an ALU, control unit, and embedded registers creates a *central processing unit* (CPU). However, the necessity to process control information for each elementary operation obviously requires additional time, energy, and hardware resources. All these can be considered as timing, power, and hardware overheads to the data-processing task.

Two reasons for limited performance can be associated with procedural programming. The first is that execution of *nonprimitive operations* is organized as multiple cycles of consequent primitive operations. For example, multiplication of 16-bit integers requires 16 consequent "shift" and "add" elementary operations in the ALU. Because of this, the execution of nonprimitive operations takes much more time than elementary operations, which reduces performance.

The second reason is that execution of *massive data* (e.g., arrays, vectors, matrices) is usually organized in loops that require a procedural overhead. This overhead includes pointers for data-elements in arrays or vectors, loop counters, pointer and counter increment or decrement operations, comparison, and branch instructions. All these are service operations with associated execution times that reduce CSPP performance as they accumulate.

To illustrate CSPP performance and the aforementioned limitations, an example of data-array execution is presented later. The following function should be executed:

$$Y = \sum_{i=0}^{1023} (k * A_i + s * B_i), \quad \text{where } i = 0, 1, \dots, 1023$$

and data-words in arrays A and B as well as coefficients k and s are 16-bit words.

The assembly language type, pseudocode, for the aforementioned function (not associated with any specific instruction set) may look as follows:

1. *Set $i = 0$*; set pointer i equal to 0
2. *Set $R4 = 0$*; clear accumulator – result register $R4$

3. *Move k to R0*; store value of coefficient *k* in register *R0*
4. *Move s to R1*; store value of coefficient *s* in register *R1*
5. *Loop Move Ai to R2*; store value of data-element *Ai* in register *R2*
6. *Move Bi to R3*; store value of data-element *Bi* in register *R3*
7. *Mult R0 & R2*; multiply *k * Ai* and store the result in register *R2*
8. *Mult R1 & R3*; multiply *s * Bi* and store the result in register *R3*
9. *Add R2 & R3*; add (*k * Ai*) to (*s * Bi*) and store the result in register *R3*
10. *Add R3 & R4*; add the content of register *R3* to accumulator *R4*
11. *Increment i*; increment pointer $i = i + 1$
12. *If i < 1024 goto Loop*; repeat *Loop* sequence until all array elements are executed
13. *Store R4 in Memory*; store result in accumulator *R4* in memory location

Assuming that

1. Each statement in this pseudocode is equal to one instruction.
2. Each instruction requires IF, ID, DF, EXE, and SR stages for execution.
3. Each stage of instruction execution needs one clock cycle including EXE stage for all elementary arithmetic–logic and data-transfer operations.
4. EXE stage for multiplication operation of two 16-bit data-words requires 16 clock cycles (assuming that "shift" and "add" operations are done in 1 clock cycle).

The execution time or T_{exe} (in clock cycles or c.c.) of this program is equal to

$$T_{exe} \text{ (CSPP)} = 4 \text{ start-up instructions} \times 5 \text{ stages}$$

$$\times 1 \text{ c.c./stage of instruction execution}$$

$$+ 1024 \times [6 \text{ instructions with elementary operations}$$

$$\times 5 \text{ stages} \times 1 \text{ c.c./instruction stage}]$$

$$+ 1024 \times [2 \text{ instructions with multiplication operation}$$

$$\times (4 \text{ stages} \times 1 \text{ c.c.} + 16 \text{ c.c./EXE})]$$

$$+ 1 \text{ store result instruction} \times 5 \text{ stages}$$

$$\times 1 \text{ c.c./instruction stage} = 71,685 \text{ clock cycles}$$

However, the count of clock cycles spent only for arithmetic and logic operations is equal to

$$\text{ALU clock cycles} = (2\,\text{Multiplications} \times 16\,\text{c.c.} + 2\,\text{Adds} \times 1\,\text{c.c.})$$

$$\times 1024 = \underline{34{,}816\,\text{clock cycles}}$$

Thus, the instruction execution and loop organization overhead amounts to

$$(71{,}685\,\text{clock cycles} - 34{,}816\,\text{clock cycles})/34{,}816\,\text{clock cycles}$$
$$= 1.05896 \times 100\% = {\sim}106\%$$

Obviously, this overhead depends on many factors (e.g., types of operations, loop organization, data structure) but is always present for any computing system with programmable procedures (CSPP). And in many cases the aforementioned overhead can exceed 100%. In other words, the number of clock cycles spent for service operations associated with control information processing and service operations exceeds the number of clock cycles spent for actual data-execution. Therefore, the *penalty* for the relatively simple and low cost of CSPP architecture is timing and power overheads. Hence, in the cases where task execution time is critical and the CSPP cannot satisfy performance requirements, other concepts of computing architecture should be considered.

1.6 Concept of an Application-Specific Computing System

Idea: If a task is considered as one information object and architecture is optimized for the algorithm and data structure of the entire task, the highest performance could be reached.

Consideration of the task as a single information object eliminates the need to divide the algorithm into elementary operations and, thus, minimizes control information processing. In addition, natural parallelism in the algorithm of the task and its data structure can be utilized efficiently. Therefore, the performance of the system can be significantly increased.

Let us take the same example task discussed in Section 1.5 and analyze the performance acceleration due to architecture optimization:

$$Y = \sum_{i=0}^{1023} (k * A_i + s * B_i), \quad \text{where } i = 0,1,\ldots,1023$$

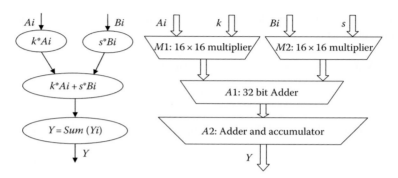

FIGURE 1.1
Task dataflow graph and architecture of the application-specific processor.

and data-words in arrays A and B as well as coefficients k and s are 16-bit words. The structure of this algorithm is shown in a form of a dataflow graph (DFG) in Figure 1.1. The DFG is a graph that represents data dependencies between operations. These operations may not be elementary and are ordered according to algorithm structure. Thus, operations are associated with DFG vertices and DFG edges show the data dependencies between operations.

If for each operation in the DFG task the appropriate hardware operational unit is assigned and information links follow the DFG edges, an application-specific architecture can be created accordingly. This architecture is shown in Figure 1.1. Often such systems are called application-specific processors (ASPs). Assuming the same timing for each operation as in the example discussed in Section 1.5 (1 clock cycle for "add" operation and 16 clock cycles for "multiplication" operation), it is possible to calculate the execution time for this task:

$$T_{exe}(ASP) = Latency + cycle\ time * (1024 - 1)$$

Here, *latency* is the period of time from the start of processing to the moment when the first result (Y_0) is stored in the accumulator. Cycle time is the period of time required for generating subsequent results—Y_i, $i = 1, 2,..., 1023$. In the case of a pipelined data-path, the cycle time is equal to the maximum operating time of each operational unit (e.g., multipliers M_1, M_2, A_1, and A_2).

Thus,

$$Latency = 16\ c.c.\ (for\ M_1\ and\ M_2\ parallel\ work) + 1\ c.c.\ (for\ A_1)$$
$$+ 1\ c.c.\ (for\ A_2) = 18\ c.c.$$

$$Cycle\ time = max\ \{16\ c.c.\ (for\ M_1\ and\ M_2\ parallel\ work);$$
$$1\ c.c.\ (for\ A_1);\ 1\ c.c.\ (for\ A_2)\} = 16\ c.c.$$

$$T_{exe}(\text{ASP}) = 18 \text{ c.c.} + 1023 * 16 \text{ c.c.} = 16{,}386 \text{ c.c.}$$

The speedup of this data-path comparing to CSPP is equal to

$$T_{exe}(\text{CSPP}) / T_{exe}(\text{ASP}) = 71{,}685 \text{ c.c.} / 16{,}386 \text{ c.c.} = 4.3748 * 100\% = 437.48\%$$

The reasons for this speedup are the following:

1. There is no need to process any control information (e.g., instructions) because for each operation a dedicated operational unit is assigned.
2. There is no need to retrieve the source operands and store the result because information transfer is organized between operational units via dedicated links optimized for the algorithm and data structure.
3. The natural parallelism of the algorithm can be utilized (e.g., parallel data-processing in multipliers M_1 and M_2).
4. Temporal parallelism (pipelining) in the data-path can be implemented.

Furthermore, the clock rate for ASPs can be much higher than for CSPPs. In addition, dedicated operational (functional) units can have higher performance than those used in a simple ALU. In the example earlier, multiplication in the ALU required 16 cycles of "shift" and "add" operations for 16 bit × 16 bit unsigned multiplication. However, high-performance multipliers utilizing Booth's algorithm [11] and pipelined multiplication [12] can also be used, which increase the throughput of the multiplier. In turn, the performance of the ASP on a given task can be higher than the performance of a general-purpose CSPP from one to two orders of magnitude.

It is necessary to mention that modern architectures of programmable processors include function-specific operational units (e.g., integer and floating-point multipliers/dividers, floating-point adders) that accelerate the operational part of instruction execution. However, all other stages of instruction execution still must be processed [13].

The ASP class of computing systems has, however, its own drawback. This drawback is associated with the complexity of ASP design and implementation. Being once designed, the architecture of an ASP is implemented in the form of an ASIC. In this architecture everything is fixed: components, links, and the functionality of both:

$$A_{asp} = \underbrace{\{C_i, L_{ij}, P_{ij}\}}_{\text{fixed}}, \quad \text{where } i, j = 1, 2, \ldots, N$$

Another issue with this approach is the much longer time for ASP design, implementation, test, and functional verification. All of the aforementioned conditions lead to a dramatic increase in design cost. Therefore, the ASP approach in form of ASIC is cost-effective in the case of mass production when the design and implementation costs are distributed between hundreds of thousand or even millions of devices.

1.7 Concept of the Computing System with Programmable Architecture

In previous sections, two extreme variants of architectural paradigms have been considered:

1. A paradigm that proposes the use of a universal computing system based on a relatively cheap hardware implementation with a relatively short application-to-architecture adaptation time through procedural programming. This paradigm has some conceptual performance limitations and requires sufficient overheads in hardware resources, time, and power for control information processing.
2. A paradigm that allows maximum performance characteristics. However, the fixed architecture is optimized for the particular task and, thus, does not provide any flexibility for rapid adjustment to other tasks or task modifications.

It is obvious that some intermediate solution(s) should be found to fill the performance *gap* between the two extremes mentioned earlier and to provide more flexibility for high-performance computing platforms.

Since variation of procedures was already used in CSPP, there are two other parts of the architecture to be considered as variable—components and links between components. Therefore, the most flexible computing architecture will be the architecture where *all three parts*, that is, components, links between components, and procedures, are variable. Thus, computing systems where (1) types of components and their functionalities, (2) topology of links between components, and (3) interface protocols and data-execution/data storage procedures are reprogrammable can be defined as computing systems with programmable architecture or *RCSs*.

Definition 1.3 *RCS* is the system where (1) functional components, (2) links between components, and (3) functional procedures are variable in time:

$$A_{rcs} = \underbrace{\{C_i, L_{ij}, P_{ij}\}}_{\text{variable}}, \quad \text{where } i, j = 1, 2, \ldots, N$$

The paradigm of RCS can combine the flexibility of CSPP with the performance acceleration provided by the ASP paradigm. This combination allows

RCS avoiding performance limitations and timing/power overheads associated with control information processing. Indeed, any task or task segment can be implemented as an ASP circuit in RCS. However, this ASP will not remain in a system forever but can be reprogrammed into another circuit when the task execution is completed.

In addition to the aforementioned, it is possible to add a set of performance constraints (e.g., response time, data rate or data throughput, power consumption limit) for each task or even task mode. Those constraints can vary depending on workload or environmental requirements. Hence, for each task or task segment, it is possible to find the optimal data-processing circuit architecture that will satisfy the set of current constraints and reprogram the system to that architecture when necessary:

$$\left.\begin{array}{ll} \text{Task segment} & TS_i \\ \text{Data structure} & DS_i \\ \text{Set of constraints} & SC_i \end{array}\right\} \text{Application specific processing circuit} \\ \rightarrow \text{ASP}_i = \{C_i, L_i, P_i\}$$

The existence of a set of constraints in the ASP architecture (e.g., available power or area) dictates the necessity of having multiple ASP architectures for the same task algorithm and data structure for cases where constraints are variable in time. This allows the reprogramming of the RCS architecture in case of changes in performance requirements or environmental conditions. That makes RCS adaptive to performance and environmental conditions and, thus, adds a new dimension of effectiveness to such systems when compared to CSPP or ASP paradigms.

The downside of the RCS paradigm is certain hardware and timing overhead that is necessary for the reconfiguration of RCS resources. The organization of RCS, including all major architectural components, will be considered in the next section.

1.8 Organization of RCS and Major Components of the RCS Architecture

The ability for reconfiguration in any system requires the existence of an array(s) of identical elements in its architecture. For example, in CSPPs, the program memory consists of identical memory cells that can be loaded with the required procedure. Each cell in the program memory contains information associated with elementary operations and data structures—*instruction*. Therefore, in CSPP architectures, the *array of identical cells* of the program memory is enough for the adaptation of the CSPP architecture to tasks.

Since the RCS concept assumes reconfigurability of all parts of the architecture, there should be three arrays of identical (1) elementary PEs as components with programmable functionality, (2) elementary communication elements (routing lines, switches, switching matrices, and input/output (I/O) interface blocks) as links with programmable topology of interconnection, and (3) elementary memory elements (MEs) as storages for programmable functional procedures. These homogenous elements can be combined into *FCR*.

Definition 1.4 *FCR* is the set of arrays of identical data-processing, communication, and memory elements with programmability of their functions, links topology, and memory configuration.

The FCR should be considered in all levels of the system:

1. On-chip level: as system-on-programmable chip
2. Onboard level: as system on board (SoB)
3. System level that may assume SoB or multiboard system organization

Since the system architecture should be programmed in the FCR prior to data-processing, a special memory is necessary to store the information regarding the required functionality components, topology of links, and data-processing and communication procedures.

Definition 1.5 *Configuration memory* is the storage for architecture configuration information.

According to configuration information associated with on-chip or system levels of the FCR, the configuration memory also can be considered as the on-chip (internal) or system-level (external) configuration memory.

Definition 1.6 *Configuration bit-file* is the file of information necessary for architecture programming of any part (components) or the entire computing system.

Configuration files can be associated with on-chip or system levels according to allocation of the respective parts of architecture to be programmed. The configuration files often were called *configuration bit-streams* reflecting streamed nature of such files. However, in further chapters the term *configuration bit-file* will be used.

In addition to this, an RCS needs some mechanism for loading/reloading bit-files from a *dedicated storage system* into the configuration memory. This loading process should be done at specific periods of time (e.g., start-up time, mode-switching time). The configuration bit-files' loading process should also be synchronized with data-execution processes (e.g., data-execution interruption or system suspension) to avoid disturbance in data-execution. In addition, the loading or reloading process of configuration bit-file(s) may be required to complete within limited period(s) of time. All of the aforementioned dictate the necessity for reconfiguration management mechanism in RCS—*resource configuration controller (RCC)*.

It is necessary to mention that all parts of the RCS architecture except the RCC are *homogenous*—consisting of identical elements. This is true for the data-execution part that consists of identical PEs, the information storage part that consists of identical MEs, the data-communication part including identical communication lines and switches and I/O interface elements, the field of configuration memory, and the storage of configuration bit streams. The general organization of the RCS is shown in Figure 1.2.

However, resource, timing, and power overheads still exist in the RCS architecture:

1. *The resource overhead* consists of static and dynamic parts. The following hardware elements can be considered a static resource overhead (i.e., always present in the system): (a) the field of configuration memory, (b) the storage of configuration bit-files, and (c) the resource configuration controller. In addition, dynamic resource overhead can exist in the system depending on task specifics and the allocation of resources for this task. This overhead consists of unused routing, PE and ME, and I/O elements in the FCR.

2. *The timing overhead* consists of (a) time for resource configuration and (b) time for synchronization and start-up of new computing circuits in the system.

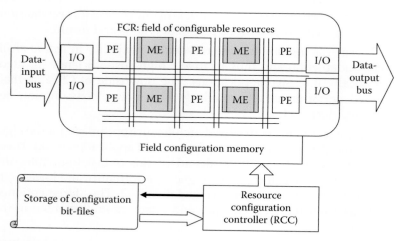

FIGURE 1.2
General architecture of the reconfigurable computing system.

3. *The power overhead* also consists of static and dynamic parts. The static power overhead includes extra static power for all elements of the hardware overhead mentioned earlier. The dynamic power overhead is a combination of the extra power spent for configuration processes (loading the field of configuration memory with configuration bit-files) and the dynamic power consumption of unused PE, ME, and other unused hardware resources.

The aforementioned overheads, being additional hardware and power resources, reduce the cost-efficiency and power efficiency of RCS in comparison to ASIC-based systems but provide significant flexibility and run-time adaptability keeping the performance characteristics close to ASIC-based systems.

1.9 Summary

Summarizing the aforementioned, it is possible to say that from the most general point of view, all computing systems can be divided into three categories:

1. CSPP, where hardware components as well as their information links are fixed. Only procedures associated with data acquisition, data-execution, and data storage are programmable. The programming in these systems is done on the basis of instructions from the given instruction set. Each instruction in CSPP class of systems represents an *elementary task* that consists of one *elementary operation* and elementary *data structure* (one pair of operands and result).

   These systems can provide reasonable flexibility in *task-to-architecture* adaptation and thus relatively high cost-efficiency for algorithmically intensive applications. The downside of CSPP is their limited performance and relatively high power consumption per executed data-unit.

2. Computing systems with a fixed architecture optimized for the algorithm and data structure of one determined application (task). These systems have no flexibility in the adaptation of task-to-architecture or vice-versa, but can reach the highest performance and lowest rate of power consumption per executed data-unit. The aforementioned systems are usually called ASPs and often are implemented in the form of ASICs.

3. RCSs or computing systems with programmable architecture, which are positioned between the first two classes of computing

systems. Being flexible in all parts of its architecture—components, information links between components, and procedures—an RCS can reach performance characteristics close to an ASP. At the same time, programmability of the RCS architecture makes possible rapid adaptation to any application by loading configuration files to the configuration memory of the RCS. The *cost* of such flexibility is certain hardware and power overheads as well as additional time for architecture reprogramming.

The classification diagram and associated table of pros and cons for each class of computing systems are presented in Figure 1.3.

Based on Figure 1.3, it is possible to see that RCSs fill the gap between CSPPs and application-specific (dedicated) computing systems. Thus, it is important to mention that the RCS paradigm does not replace either of the other two computing paradigms, but is complementary to them. Furthermore, it is possible to see that CSPP or ASP paradigms are particular cases (subclasses) of the RCS paradigm.

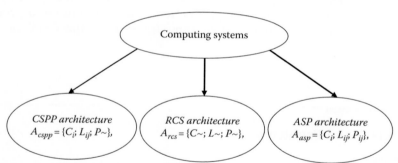

Class of computing system	Computing systems with programmable procedure (CSPP)	Computing systems with programmable architecture—reconfigurable computing systems (RCS)	Computing systems with dedicated architecture—application-specific processors (ASP)
Flexibility	Moderate	Highest	Limited
Performance	Limited	High	Highest
Power efficiency	Low	High	Highest
Cost in mass production	Relatively low	Relatively high	Relatively low
System design time	Relatively short	Moderate	Relatively long
Best class of applications	Algorithmic-intensive application	Computation-intensive applications	Computation-intensive applications

FIGURE 1.3
Classification of computing systems according to flexibility of architecture.

Utilization of RCS is effective in areas associated with

1. Ability for real-time adaptation to environmental/workload changes or constraint changes associated with mode of operation— *multimodal applications.*

2. Massive data-processing in limited time: DSP, multimedia, video and image processing, data mining, digital communication and broadcasting, and many others—*computation-intensive applications.*

3. Multitask applications where several independent tasks are composing the workload and any combination of these tasks can run in parallel being initiated independently.

4. Highly dependable and self-healing systems that can provide self-recovery in cases of hardware faults. This includes systems that can provide gradual performance degradation in case of permanent faults of processing communication or MEs.

5. Prototyping of ASPs for further implementation as ASICs.

In the aforementioned applications, utilization of RCS paradigm may not be effective in the case of mass production. The best volume of production should always be analyzed in comparison with the CSPP or ASP approaches.

Exercises and Problems

Exercises

1. Why was the concept of reconfigurable computing systems called "antimachine"?

2. Define "computing architecture" in formal terms of components, links, and functional procedures.

3. Determine the possible classes of computing paradigms according to variability of each element composing computing architecture.

4. List the forms of component implementation in computing system with programmable procedure and in the computing system with programmable architecture.

5. Why can the application-specific processor, being implemented in the form of ASIC, reach the highest data-processing rate and the best power efficiency in comparison to conventional computers with programmable procedure and RCS? List the reasons.

6. For which applications is the concept of CSPP the most suitable? List the reasons why.

7. For which applications is the concept of ASP the most suitable? List the reasons why.
8. For which applications is the concept of RCS the most suitable? List the reasons why.

Problem

For video/image processing tasks the YCbCr color conversion is one of the most utilized procedures. The gamma-adjusted *luma* component (Y') can be calculated according to the gamma-adjusted *red* (R'), *green* (G'), and *blue* (B') values of each pixel using the following formula [14]:

$$Y'_i = 16 + (K_r * R'_i + K_g * G'_i + K_b * B'_i) / 256, \quad \text{where } i \text{ is pixel number}$$

Assuming that K_r, K_g, K_b, and RGB pixel values are 8-bit integers, *estimate the Y' execution time (in clock cycles)* for all pixels in the video graphics array (VGA) video-frame ($640 \times 480 = 307{,}200$ pixels) in case of implementation on the conventional instruction processor (class CSPP) and in case of implementation in ASIC-like ASP. The estimation can be done similar to examples discussed in Sections 1.4 and 1.5. Determine the speedup of ASP assuming that the cycle time for the fully pipelined ASP can be reduced up to one clock cycle.

References

1. J. von Neumann, First draft of a report on the EDVAC, Contract #.W-670-ORD-4926, Moore School of Electrical Engineering, University of Pennsylvania, Philadelphia, PA, June 30, 1945.
2. G. Estrin, Organization of computer systems—The fixed plus variable structure computer, in *Proceedings of the Western Joint Computer Conference*, New York, 1960, pp. 33–40.
3. G. Estrin, B. Bussell, R. Turn, and J. Bibb, Parallel processing in a restructurable computer system, *IEEE Transactions on Electronic Computers*, 12(6), 747–754, December 1963.
4. R. Hartenstein, Reconfigurable computing and the von Neumann syndrome (opening keynote), in *Seventh International Conference on Algorithms and Architectures for Parallel Processing (ICA3PP)*, Hangzhou, China, June 11–14, 2007.
5. R. Hartenstein, A decade of reconfigurable computing: A visionary retrospective, in *DATE-2001: Proceedings of the Conference on Design, Automation and Test*, Munich, Germany, 2001, pp. 642–649.
6. M. J. Flynn, Some computer organizations and their effectiveness, *IEEE Transactions on Computers*, 21(9), 948–960, September 1972.
7. J. Elke, Origin of the virtual memory concept, "https://en.wikipedia.org/wiki/IEEE_Annals_of_the_History_of_Computing" \o "IEEE Annals of the History of Computing" IEEE Annals of the History of Computing 26(4), 71–72, 2004.

8. A. M. Turing, On computable numbers, with an application to the Entscheidungs problem, *Proceedings of the London Mathematical Society*, 2(42), 230–265, 1937.
9. Z. Konrad, *The Computer—My Life*. Springer-Verlag, Berlin/Heidelberg, 1993.
10. D. A. Patterson and J. L. Hennessy, *Computer Organization and Design*, 4th ed., Morgan Kaufmann is an imprint of Elsevier Inc., Waltham, MA, 2012, Chapter 4, pp. 300–330.
11. W.-C. Yeh and C.-W. Jen, High speed booth encoded parallel multiplier design, *IEEE Transactions on Computers*, 49(7), 692–701, July 2000.
12. D. A. Patterson and J. L. Hennessy, *Computer Organization and Design*, 3rd ed., Morgan Kaufmann is an imprint of Elsevier Inc., San Francisco, CA, 2005, Chapter 3, p. 184.
13. D. A. Patterson and J. L. Hennessy, *Computer Architecture: A Quantitative Approach*, 5th ed., Morgan Kaufmann is an imprint of Elsevier Inc., Waltham, MA, 2012, Chapter 3, pp. 148–192.
14. C. Poynton, *Digital Video and HD: Algorithms and Interfaces*, 2nd ed., Morgan Kaufmann is an imprint of Elsevier Inc., Waltham, MA, January 2012.

2

Organization of the Field of Configurable Resources

2.1 Introduction

In Chapter 1, the generic organization of the reconfigurable computing system (RCS) was discussed, and the definition of the field of configurable resources (FCR) was given according to the concept of computing systems with programmable architecture. In this regard, two fundamental aspects must be considered when describing an FCR: (1) the granularity of the basic elements of the FCR and (2) the organization of the synthesis of larger components from basic elements. The concept of component granularity and homogenous versus heterogeneous organization of the FCR will be considered in this chapter.

Following this, the static versus dynamic synthesis of processing circuits and the partitioning of FCR into task segments will be discussed.

The two aspects described earlier are the most fundamental and general aspects to be considered when discussing any paradigm associated with architecture synthesis from uniform components. These aspects can be seen in many areas, including city architecture, societal organization, and life-form evolution. They lead to the following three main questions:

1. What is the nature and granularity of the building blocks of architecture?

2. How homogenous or heterogeneous is the field of these functional resources?

3. What is the synthesis mechanism (static or dynamic) that allows generation of more complex structures using these building blocks?

The answers to these questions have direct influence on the flexibility and, thus, adaptability of the system architecture and, as a result, on the effectiveness and survivability of the system.

2.2 Granularity of Logic Elements for the FCR

As mentioned in Section 1.3, a "component" can be defined as an *information object* that is being deployed on a platform, performs certain function(s), and communicates with other components or the external world via a specific interface. Hence, the component by itself is not considered a hardware, software, or firmware unit but just an information object. This information object can be represented in many forms: as text, as a behavioral model, in symbolic form, or in other graphical forms. However, regardless of the method of representation, implementation of a component requires some hardware element(s). In digital systems, these hardware elements are called logic elements assuming that all functions can be implemented in logic circuits. Here appears the dilemma: *granularity of logic elements versus connectivity between them.*

2.2.1 Granularity of Basic Logic Elements versus Complexity of Connectivity

Granularity of logic elements is one of the major factors that influence the performance characteristics of the system and its cost efficiency.

Definition 2.1 *Basic logic element* (*BLE*) is the smallest hardware element used for the composition of any components in the system that is able to process and/or store the data.

The benefit of smaller logic elements is the same as with any small "bricks" in the construction of large buildings—the algorithm and data structure of a task can be implemented in hardware elements more precisely. This leads to an increase in efficiency of hardware utilization for all tasks in the workload and, therefore, better performance per cost.

However, there are certain drawbacks of this approach. If a relatively small logic element is taken as the basic element, a more complex network of interconnections should be deployed in FCR to compose the functional component or the entire system. A complex network, in turn, requires more pass transistors or switching elements to create the set of links between logic elements of a component or system, which leads to

1. Higher hardware overhead for components or system configuration
2. A more complicated compilation process associated with place-and-route (PAR) procedures
3. A higher propagation delay for data-transfer and, thus, lower system performance
4. More configuration bits to be stored in configuration memory for component's or system architecture programming and, therefore,

(a) the need for a higher volume of configuration memory and (b) a longer FCR configuration time

Many of the earlier aspects associated with PAR procedures, timing closure, and physical synthesis of very large scale integration (VLSI) circuits are analyzed in detail in [1]. A large volume of research publications exist regarding algorithms and methods for PAR and low-level synthesis of processing circuits in field-programmable gate array (FPGA) devices (e.g., [2,3]).

On the other hand, utilization of a *larger BLE* in the FCR assumes more built-in programmable functions in each element. This, in turn, simplifies the communication network, which allows the system to work with higher clock frequencies. Indeed, a larger data-execution element means that more information exchange is localized in the processing element and, thus, less information needs to be transferred between elements via the configurable communication network. In such an organization, local communication lines (inside the element) are static by its nature, relatively short, and usually dedicated (peer to peer, without pass transistors). Therefore, propagation delays in these lines are much lower compared with propagation delays in relatively long interelement communication lines. Hence, the operational frequency in such elements can be higher as well as associated performance.

At the same time, interelement communication links can be simplified, assuming there are less switching elements between communication points (ports). This aspect was well described in [4] where several RCS architectures based on *coarse-grained elements* were analyzed.

A similar approach can be found in the difference between highways and regular streets in cities. The network of regular streets has a lot of crossings and traffic lights (road switches), thus limiting speed. However, streets are relatively short in distance, and thus, travel time is short accordingly. In contrast, highways are much longer than streets but allow higher speed by excluding road crossings. This allows minimization of travel time overhead for long distances.

Looking to the city as a sort of FCR, it is possible to see that finding the proper balance between the organization of road network and set of highways is not a simple task. The same problem of balancing information traffic appears in any FCR, regardless of the nature of the resources used.

Nonetheless, in addition to higher operational frequency, utilization of larger basic elements in FCR brings another important advantage for RCS. It leads to a reduction of the configuration data-file volume. This is because less transaction switches have to be configured in the network for data-transfer between basic elements. The smaller volume of the configuration data file requires a smaller volume of configuration memory. Furthermore, a smaller configuration file volume reduces reconfiguration time in the FCR, which improves its event/mode response time.

On the other hand, a larger granularity of basic elements in FCR reduces the flexibility of the RCS and, therefore, its cost-effectiveness in adaptation to the application.

2.2.2 Granularity of Basic Logic Elements for the FCR

Out of many possible options, only three have been considered practically acting as basic information processing/storage elements: (1) logic gates, (2) configurable logic blocks (CLBs), and (3) function processing unit (FPU). The FPU as a basic element can be implemented in two forms: (1) function-specific processing unit, which performs dedicated function (e.g., floating-point multiplication or matrix multiplication), or (2) programmable FPU, which performs one of possible functions according to a programmable procedure (program) or by selection of one of possible functions from a predesigned set. In both cases, the term "function" assumes a set of elementary operations executed sequentially or in parallel.

Let us first consider *logic gates* acting as BLE [5]. It is clear that any digital computing, communication, control, or memory circuit can be designed and built using logic gates as the building blocks. This is the traditional approach, historically first, and the most general. This approach provides precise implementation of any functional component using a logic gate–based hardware platform similar to the platforms used for any integrated circuit. However, if reconfiguration is considered at the logic gate level, the number of links between logic gates is expected to be very large even for moderately complex functional components. Thus, the configuration file and configuration memory volume can be very large too. Since configuration memory is considered as hardware overhead to the FCR, this approach may not be effective for most of practical applications. Indeed, if the number of transistors utilized for configuration memory and routing switches is comparable (or even greater) than the number of transistors in logic gates, the hardware overhead can exceed 50%. Therefore, this approach, usual for the implementation of application-specific integrated circuits (ASICs), did not get much acceptance for the organization of the FCR. It was mostly used for (1) relatively small programmable logic devices (PLDs) or (2) one-time PLDs where one cycle of circuit configuration is needed and, thus, the static random access memory (SRAM)-based configuration memory does not exist [5]. Instead, a fuse (or antifuse)-based configuration solution is usually used to minimize the cost of the hardware overhead. The larger BLE to be considered is CLB.

Definition 2.2 CLB is the BLE that can be configured or programmed to perform a 1-bit-wide arithmetic/logic and bit-storage operation.

The concept of CLB in different forms was proposed and developed by many researchers and designers from the early 1970s to the end of the 1980s (e.g., [6,7]).

In its conventional form, the 1-bit-wide arithmetic–logic unit (ALU) and memory element (ME) consist of (1) the programmable logic block (PLB), which can perform bit-wide logic functions; (2) 1-bit arithmetic unit (AU) (e.g., 1-bit adder and carry-in/out nets); (3) Multiplexor (MUX), 1-bit result multiplexor; and (4) 1-bit ME. The general organization of the CLB is shown in Figure 2.1.

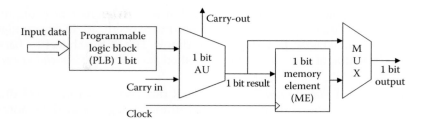

FIGURE 2.1
General organization of the configurable logic block.

The idea of CLB organization is the same as in a regular ALU—any arithmetic/logic operation can be represented as a combination of elementary operations: (1) logic AND, OR, and NOT, (2) right or left shifts, and (3) arithmetic ADD.

Therefore, a wider (larger than 1 bit) custom ALU circuit can be created from a combination of the aforementioned CLBs organized as an array. For example, a 36-bit adder can be built using at least 36 CLBs. In reality, more than 36 CLBs can be used due to the required service logic circuits. Furthermore, each of these custom ALUs could be optimized for a predetermined operation and put in sequence with other ALUs in a pipeline. This allows a significant increase in performance because execution of service functions (e.g., instruction fetch, operation decode, input data delivery, and result storage) could be excluded from the computation process.

The organization of the PLB can take one of two forms:

1. As a collection of logic elements with programmable interconnections
2. As a look-up table (LUT)-based array of configuration memory cells

The first approach was used in field-programmable logic array [8] proposed by N. Cavlan in 1981. This approach is used mostly in complex programmable logic devices (CPLDs) where AND, OR, and NOT logic gates were grouped in so-called macrocells. It could be found in Xilinx CPLDs (e.g., XC95XX series of CPLDs [9]). The second approach was used in BLEs of FPGA devices [10].

The CLB programmability is based on the configuration memory, where a certain portion of this memory is dedicated to each CLB. Thus, one of the large set of logic functions could be programmed into each CLB individually, as well as AU activation and memory unit utilization.

This aspect will be discussed in detail in Section 3.2.

In both of the cases discussed earlier, the PLB and the entire CLB are dedicated circuits. Hence, the configuration process in the FCR consists of (1) CLB programming and (2) synthesis of custom processors created using combinations of arrays of CLBs. This, obviously, requires less communication lines and switches for routing between CLBs when compared with architectures based on simple logic gate arrays. Therefore, this allows a reduction of the configuration memory volume and the associated hardware overhead. RCS reconfiguration time can, likewise, be reduced accordingly. The RCS architecture based on FCR consisting of CLBs as the BLEs is called *fine-grained RCS architectures.*

Definition 2.3 *A fine-grained FCR* is an FCR that consists of identical BLEs oriented toward 1-bit-wide elementary arithmetic, logic, and data-storage operations.

In contrast to the fine-grained architecture of RCS, the architecture is called "coarse grained" if it consists of the FPUs [11].

Definition 2.4 FPU is any processing circuit that can perform one or a set of functions, each of which consists of multiple elementary operations composed of a function-specific algorithm and associated with a determined data structure.

Definition 2.5 *A coarse-grained FCR* is an FCR that consists of identical FPUs as basic data-processing elements oriented toward multibit (word-wide) function-specific data-processing and/or data-storage operations.

The organization of FPU strongly depends on the class of application that the RCS is targeting. Therefore, there are many approaches in determining the FPU structure in the FCR and how this field should be organized. Due to their coarse-grained granularity, such FCRs have been called "coarse-grained reconfigurable arrays" [12]. The FPU can take a wide range of forms. It can vary from relatively small *silicon objects* in the field-programmable object array to relatively large programmable tiles, which include reduced instruction set computer (RISC) processing core and coprocessor for floating-point operations [12]. There can be macrofunction-oriented FPUs such as multiply–accumulate (MAC) units or CRC generators. In all cases, FPU organization reflects the specific nature of the application. In other words, if the application requires certain functions to be executed often, those functions may be embedded into appropriate hardware circuits, since their configuration is not needed. Thus, the presence of *static macro-operations* in the task algorithm leads to effective utilization of *static macroprocessing circuits* optimized for these operations. The advantage of using FPUs in the FCR of the RCS is a further simplification of the communication infrastructure and, thus, reduction of reconfiguration time. Thus, adaptation of a coarse-grained FCR to changes in workload or environment can be done more quickly when compared to fine-grained FCR. In other words, the response time of coarse-grained RCS is shorter than that of other RCS organizations.

   The disadvantage of coarse-grained FCR is the strong dependence of the RCS efficiency on the class of applications being implemented.

As a summary, it is possible to conclude that *granularity of the workload* dictates the *granularity of basic elements* in the FCR and vice versa. In other words, if the task algorithm requires many standard macro-operations (e.g., floating-point arithmetic or digital signal processing [DSP] functions), utilization of the FPUs reflecting these macro-operations can dramatically increase the effectiveness of the RCS. In contrast, if FCR contains certain types of FPUs in its structure, it would be more efficient for task algorithm implementation to use the macro-operators associated with these FPUs. Furthermore, the use of the aforementioned macro-operators in task programming simplifies the programming process and reduces task implementation time.

Essentially, a higher granularity of basic data-processing elements in FCR minimizes the volume of configuration data, configuration memory, communication routing infrastructure, and associated hardware overhead. On the other hand, a higher granularity of basic data-processing elements reduces the flexibility of the RCS and, therefore, its effectiveness at *covering* the operators and data structures used by the algorithm. As a result, there may be an imbalance between the number of available resources and the needs of the task. Hence, some resources may stay unused for certain periods of time, or alternatively, there may be delays in data-processing due to the availability of resources for task operations. Smaller FCR granularity allows for more precise utilization of hardware resources for different operations in the task but requires a more complex communication infrastructure and a bigger volume of configuration memory as well as reconfiguration time. This is one of the major trade-offs associated with the organization of the FCR in RCS.

2.3 Heterogeneous Organization of the FCR

In the previous section, we have considered an FCR organized as an array of identical basic elements and, thus, homogenous in its nature. However, each level of granularity of such an FCR was found to offer certain advantages and drawbacks. Thus, a mixed FCR organization could compensate the shortcomings of one type of basic element (e.g., fine grained) with benefits provided by another (e.g., coarse grained).

2.3.1 General Organization of the Heterogeneous FCR

Based on this concept, the idea of a mixed FCR organization appears to be productive. Such an idea of combining different basic elements in the same FCR to compensate the negative aspects of each type of element is usual and can be found in the organization of a number of systems. For example, on sailboats the negative aspect of the sail as a propelling mechanism is its dependence on wind and its direction. This negative effect can be compensated by the inclusion of a combustion engine as an additional propelling system to the sailboat. And vice versa, the negative aspects of the combustion engine (fuel consumption and noise) can be compensated by a sail, which allows wind power–based travel. In the case of conventional computing systems, the central processing unit (CPU) optimized for integer scalar operations is often combined with a coprocessor optimized for floating-point operations. Systems that combine different types of basic elements in their organization, with the aim of compensating drawbacks of each type of elements, are referred to as heterogeneous systems.

Definition 2.6 *A heterogeneous FCR* is an FCR that consists of several types of basic elements with different granularity and functional specification.

In the case of FCR organization, the idea of heterogeneous organization means using a combination of elements with different levels of granularity and different functionality to provide the best set of resources for certain classes of applications [13].

In practice, it means that an FCR should be organized as a combination of the following arrays:

1. Array of CLBs
2. Array of FPUs
3. Arrays of functional supporting blocks (memory modules, interface units, etc.)

The FPUs in such an organization are usually application specific and are optimized for acceleration of execution of macro-operations common to certain classes of applications. For instance, FPUs for DSP applications can include an array of MAC units [14]. For applications that require intensive data communication, FCR can contain an array of serializer/deserializer (SerDes) units. In addition, the arrays of functional supporting blocks may include arrays of memory blocks, arrays of interface (input/output) blocks, and a set of clock frequency management blocks (including delay locked loop [DLL] or phase locked loop [PLL] circuits).

In cases where the application consists of both algorithmic and computation-intensive segments, hard-core RISC processors with associated memory blocks and busses can also be included in the FCR. A general organization of a heterogeneous FCR is shown in Figure 2.2. A more detailed discussion regarding implementations of heterogeneous FCR will be considered in Section 3.3.

As shown in Figure 2.2, in general organization of heterogeneous FCR, it consists of several types of configurable resources—R_i, where $i = 1, 2,...,n$. Each type of resource consists of multiple identical instances (elements) composed in array (1D or 2D).

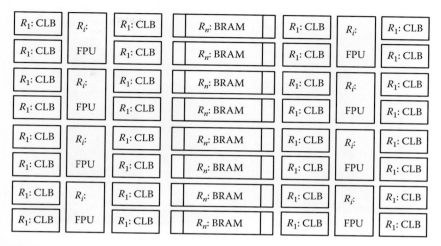

FIGURE 2.2
General organization of a heterogeneous field of configurable resources in a reconfigurable computing system.

2.3.2 Multilevel Organization of Heterogeneous FCR

It is necessary to mention that FCR resources should be considered at all levels:

1. *On-chip level*: This level of the FCR assumes configurable resources deployed inside the chip that can provide the ability to program the hardware architecture inside the chip (e.g., FPGA device and/or CPLD).

2. *On-board level*: This level of the FCR assumes configurable resources deployed on the board (e.g., array of FPGA devices or CPLDs) with associated support circuits.

3. *System level*: This level of the FCR assumes all configurable resources in the system (e.g., the array of boards with FPGA devices or CPLDs).

In other words, the FCR in a system is not limited to on-chip configurable resources only. The FCR should be considered at several levels and can consist of multiple PLDs on a board or even multiple boards with reconfigurable logic devices in the entire RCS. The closest analogy for the FCR concept in conventional computing systems is the memory hierarchy. Each memory module or device consists of identical data-storage elements and thus seems homogenous. However, data access patterns lead to memory division into cache, main memory, and secondary storages (e.g., hard disc drives, flash cards). As a result, the complete memory hierarchy becomes more heterogeneous in nature. The complete hierarchy consists of three levels: (1) on-chip

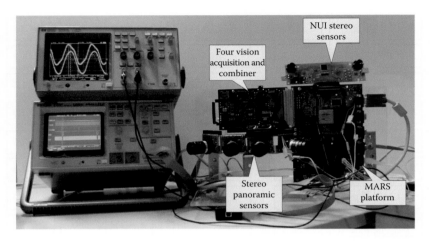

FIGURE 2.3
Multiboard reconfigurable computing system setup with multilevel field of configurable resources. (Courtesy of Embedded Reconfigurable Systems Lab (ERSL), Ryerson University, Toronto, Canada.)

level (e.g., cache SRAM or main memory DRAM), (2) memory board level (e.g., single in-line memory module or dual in-line memory module), and (3) system level including all memory devices in the system.

An example of multiboard and multilevel organization of FCR in reconfigurable computing system is depicted in Figure 2.3 (system level). The RCS consists of (1) stereo panoramic video-sensor composed of four video-sensor modules integrated with four vision image acquisition and transport stream combining module, (2) natural user interface module based on 3D image processing, and (3) multi-stream adaptive reconfigurable system.

Each of the listed modules is deployed on separate boards as shown in Figure 2.4. Each board contains the core FPGA-performing dedicated functions associated with a particular module and according to one of the possible modes of operation.

Therefore, the entire FCR of this RCS consists of several FPGA devices and CPLDs allocated on different boards. FPGAs and CPLDs deployed on the same function-specific board are representing the on-board level of FCR. All FPGAs and CPLDs consist of basic elements with different granularities: (1) fine-grained CLBs (in FPGAs) and macrocells (in CPLDs), (2) coarse-grained DSP modules (in some FPGAs), (3) block-RAM modules (in FPGAs), etc.

Furthermore, each FPGA or CPLD can be considered as coarse-grained FPU with the ability to change its functionality according to application needs.

The organization of RCS on system and on-board levels will be discussed in Chapter 4.

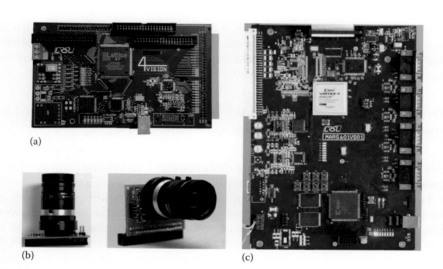

FIGURE 2.4
On-board level of the reconfigurable computing system organization: (a) Four vision acquisition and combiner, (b) video sensor modules, and (c) multi-stream adaptive reconfigurable system (MARS). (Courtesy of Embedded Reconfigurable Systems Lab (ERSL), Ryerson University, Toronto, Canada.)

2.3.3 Effectiveness of Resource Utilization in Heterogeneous FCR

After discussion of organization of the multilevel heterogeneous FCR, the effectiveness of resource utilization can be considered.

Let us assume that the total number of elements of resource type R_i is NR_i. If the area used for one R_i element is equal to a_i, then the total area used for all R_i resources in the FCR can be obtained by the following equation:

$$\text{Area} (R_i) = NR_i \times a_i \tag{2.1}$$

Then, the total area of all resources in FCR can be obtained by the following equation:

$$\text{Area (FCR)} = \sum_{i=1}^{n} (NR_i \times a_i) \tag{2.2}$$

Resource utilization effectiveness in FCR depends on how many elements of each resource type can be used during the period of task execution. In contrast with fixed architecture systems (e.g., ASICs), the resource utilization in FCR should be considered *in space and in time*. Indeed, if some resources are already deployed in FCR but have not been used for a certain period of time, they can be considered as hardware overhead. Furthermore, unused resources consume power and thus lead to an energy overhead as well.

Therefore, one of the goals of RCS architecture synthesis and design is minimization of hardware and power overheads during the task execution. For evaluation of RCS effectiveness for a given task, the *resource utilization factor (RUF)* can be defined.

Definition 2.7 *The RUF* represents the percentage of resources utilized in space and in time for a given task or application.

According to the value of RUF, the designer can estimate and compare the quality of FCR utilization for any application.

Let us assume for the beginning that no reconfiguration of resources is needed within the task execution period. In other words, the system behaves like a system with fixed architecture for the duration of the task. Such RCS can be defined as a *statically configurable computing system*. Since the only FCR in the RCS allows configuration, statically configurable RCS assumes statically configurable FCR (SFCR).

Definition 2.8 *An SFCR* is an FCR where configuration of all resources into computing architecture is completed prior to task execution and remains the same during the period of task execution.

For this case, the effectiveness of resource utilization does not depend on the task execution time, and thus, RUF can be determined as the following ratio:

$$\text{RUF}_{\text{SFCR}} = \frac{\sum_{i=1}^{n} (UR_i \times a_i)}{\sum_{i=1}^{n} (NR_i \times a_i)} \times 100\% \tag{2.3}$$

where UR_i, $i = 1, 2, \ldots, n$ is the number of elements of resource R_i used in the FCR for given task processing. For example, let us consider the following organization of an FCR:

1. The FCR consists of three types of resources:
   a. R_1: CLBs, each of which requires 10 area units (au)
   b. R_2: Macro-operation components (MAC), deployment of which needs 100 au
   c. R_3: Built-in random access memory (BRAM) modules that require 1000 au
2. The FCR contains 1000 CLBs, 100 MACs, and 10 BRAMs.
3. The task being executed requires 850 CLBs, 64 MACs, and 4 BRAMs.

For this task and available resources, the RUF can be calculated as follows:

$$\mathrm{RUF} = \frac{850\ \mathrm{CLBs} \times 10\ \mathrm{au} + 64\ \mathrm{MACs} \times 100\ \mathrm{au} + 4\ \mathrm{BRAM} \times 1000\ \mathrm{au}}{1000\ \mathrm{CLB} \times 10\ \mathrm{au} + 100\ \mathrm{MACs} \times 100\ \mathrm{au} + 10\ \mathrm{BRAM} \times 1000\ \mathrm{au}} \times 100\%$$

$$= 63\%$$

According to the RUF definition, it is easy to see that an increase in RUF depends primarily on the percentage of coarse-grained elements used in task execution. If, in the example earlier, all 10 BRAMs are used, the RUF will increase by 20%, up to 83%. In contrast, if all CLBs are used, the RUF will increase by 5% from 63% to 68%.

It is necessary to mention that any coarse-grained element of the FCR can be composed by fine-grained elements to perform the same functionality. However, this replacement usually requires a larger area than the coarse-grained element occupies. Furthermore, an additional portion of configuration memory will also be used in the case of the aforementioned replacement. This extra area of the FCR and additional volume of configuration memory result in higher RUF value but less efficient utilization of the coarse-grained resources. Moreover, hardware and power overheads for the replacement of a coarse-grained element by a number of fine-grained elements also increase. On the other hand, in the case of a lack of coarse-grained elements for a task, some compensation is possible by employing unused fine-grained elements. For example, in case of lack of MACs, it is possible to create additional MACs out of CLBs. Obviously, the area occupied by CLB-based MACs will be greater than the area needed for coarse-grained MACs, but the task will have enough resources to run with the requested performance.

Alternatively, it is possible to use large LUTs instead of data-processing elements. These LUTs are usually implemented using BRAMs. For example, instead of logarithm calculation on specific processing circuit, the table of logarithms can be stored in the BRAM. Thus, the logarithm search procedure may replace respective logarithm calculation and can be done within one to two clock cycles.

In both cases, the value of the RUF increases because more resources are involved in the task execution process, as opposed to the remaining unused. Thus, fine-grained elements may play the role of *compensators* in cases where not enough coarse-grained elements are available for the given task. This is in addition to their role as building blocks for custom functional components.

As a summary, it is possible to see that a heterogeneous FCR organization gives many benefits to RCS due to a higher level of RUFs and, thus, better use of FCR resources for the application needs. On the other hand, the cost-efficiency of SFCR strongly depends on the specifics of the task algorithm and data structure. The proper association of macro-operations in the task algorithm and coarse-grained elements in the FCR becomes the key aspect of the cost-efficiency of the RCS.

Obviously, variations in the task algorithm and/or data structure may significantly reduce efficiency of resource utilization or even make it impossible for further execution of the task. Thus, appropriate changes in computing architecture are necessary to mitigate algorithm/data structure variations satisfying all performance requirements. These architecture changes dictate the necessity for resource reconfiguration in FCR. In other words, a new task or task variant requires respective composition of appropriate architecture on the same FCR. As well, mitigation of environmental constraints (e.g., available power, temperature variations) can also require appropriate reconfiguration of system architecture at all levels of the RCS. In turn, this fact dictates the necessity for dynamic reconfiguration of the processing architecture in FCR. This aspect will be discussed in the next sections.

2.4 Dynamic versus Static Reconfiguration of Resources in FCR

In Section 2.3.3, the static configuration of resources in FCR was analyzed. The SFCR implicitly assumes that FCR configuration time is negligibly small when compared with task execution period of time. However, there are applications where different configurations of the FCR could be requested. The necessity for reconfiguration could be caused by request of system functionality (workload variation), adaptation to the environmental conditions (e.g., temperature, power limitation, radiation effects), and hardware faults caused by radiation, ageing, thermocycling, etc. In all the aforementioned cases, the architecture of the FCR may need to be reconfigured during the period of execution of the same application. The process of FCR reconfiguration in such cases can be performed statically or dynamically.

2.4.1 Statically Reconfigurable FCR

The reconfiguration process of the entire FCR assumes termination of the task execution process and loading of a new configuration data file to this configuration memory. Within this period of time, all resources are stalled. Hence, such an organization of the reconfiguration process in the FCR reduces the RUF for any application. If we include the task processing time, T_{exe}, and the configuration time of resources in FCR, T_{conf} into Equation 2.3 of the RUF, the definition of the RUF including the FCR configuration time will look as follows:

$$\text{RUF}_{\text{SrFCR}} = \frac{\left[\sum_{1}^{n}(UR_i \times a_i)\right] \times T_{exe}}{\left[\sum_{1}^{n}(NR_i \times a_i)\right] \times (T_{exe} + T_{conf})} \times 100\% \qquad (2.4)$$

In the case where task execution time is much greater than FCR configuration time, Equation 2.4 becomes equal to Equation 2.3. Hence, the formal definition of *statically reconfigurable* FCR can be given according to the specifics mentioned.

Definition 2.9 *A statically reconfigurable FCR* is an FCR oriented toward applications that allow stall of the data execution for the period of reconfiguration for the next mode of operation.

2.4.2 Dynamically Reconfigurable FCR

However, in many practical cases, RCS working in a real environment cannot satisfy the aforementioned requirements and, thus, cannot be implemented as statically reconfigurable computing systems. Most real-time applications usually require nonstop performance of data acquisition, control, and communication processes. These systems can permit periods of interruption in processing tasks. However, these periods are time-limited and must occur at specific points in time. This situation is usual for multimodal applications or applications where prompt reaction to different variations of environmental conditions is essential. There are many technological, communication, and medical processes that are uninterruptable by their very nature. Any interruption can cause severe damage (to the machine, the environment, or the patient involved).

In such cases, there are certain moments of time when the RCS has to adapt to (1) a new mode of operation, (2) a new set of environmental conditions, or (3) a new set of performance constraints.

This adaptation can be done by reconfiguration of the RCS architecture in run-time of the application. Thus, the *architecture reconfiguration process can be considered as the RCS adaptation mechanism* that allows the system to maintain a requested level of performance despite variations in the factors listed earlier. In most cases, the real-time execution process cannot be suspended entirely. However, some segments of the task can be interrupted for a predetermined period of time. For example, in the case of uncompressed videostream processing, there are periods of time between the video-frames when no data-processing is needed. If the application is associated with network stream processing, there are periods of time when only header information should be processed from a network packet, but not a datagram, etc. To accommodate the aforementioned classes of applications, the RCS should provide architecture reconfiguration in run-time or *dynamic reconfiguration* of resources in its FCR.

Definition 2.10 *A dynamically reconfigurable FCR* is an FCR where configuration and reconfiguration of resources can be performed at *run-time* of the task execution.

An RCS with a dynamically reconfigurable FCR is called a dynamically reconfigurable computing system (DRCS) or run-time reconfigurable (RTR) RCS [15]. Furthermore, dynamic reconfiguration may not include the entire FCR. It may be applied only to some part(s) of it. In such a case, the FCR should be organized as a *partially reconfigurable* (PR) FCR [16].

Definition 2.11 *A PR FCR* is an FCR where only a portion of resources can be reconfigured, while the rest of resources are keeping their configuration.

Partial reconfiguration is possible for both statically reconfigurable and dynamically reconfigurable FCR. The partial reconfiguration being applied to statically reconfigurable FCR allows significant reduction of the reconfiguration time for the FCR when switching from one mode to another. In this case, resources not involved in the configuration process are nonetheless stalled for the duration of the configuration period. In contrast to that, partial reconfiguration of resources in a dynamically reconfigurable FCR allows the remaining resources to perform the task execution nonstop. Obviously, dynamic reconfiguration can be done only if the FCR allows partial reconfiguration. Therefore, a dynamic partially reconfigurable (DPR) FCR is the hardware platform for DRCSs. The DPR FCR consists of *blocks of configurable resources* often called frames. These blocks of resources (e.g., blocks of logic resources, memory blocks) are the smallest portion of resources that can be entirely reconfigured. The reconfiguration process in each block is performed separately from other blocks of resources in the FCR and without suspension of operation in those resources. This option allows running reconfiguration and task processing simultaneously [16].

The dynamic partial reconfiguration of the FCR will be discussed in detail in Chapter 7.

It is necessary to mention that reconfiguration may take place at certain points in time. As was mentioned earlier, the reconfiguration of FCR in RCS is the adaptation mechanism for mitigation of external events or internal (e.g., workload) requests. In all cases, the RCS reaction to the aforementioned events is to make changes to the processing architecture by complete or partial reconfiguration.

In general case, any request for change in (1) task algorithm, (2) datastructure, or (3) operational constraints (e.g., processing data rate, power limit) can cause a change of *mode of operation*.

Definition 2.12 *A mode of operation* is one of multiple possible combinations of the task algorithm, associated data structure, and/or set of operational/ performance constraints.

Thus, the architecture adaptation to the mode of operation, M_i, can be considered as the selection and implementation of the architecture variant

$A_i = \{C_i; L_i; P_i\}$ optimized for the given combination of task algorithm (TA_i), data structure (DS_i), and set of constraints (SC_i):

$$A_i = \{C_i; L_i; P_i\} \rightarrow \{TA_i; DS_i; SC_i\}$$

There are two possible options for the implementation of the aforementioned adaptation process:

1. Compilation of the *entire configuration* of resources according to the architecture A_i optimized for each mode M_j. Then, *static reconfiguration* of the FCR for each A_i, $i = 1, 2,\ldots, m$ is implemented according to the requested mode.
2. Compilation of the basic (initial) FCR configuration and *changes in basic configuration* associated with mode variations. Then, *dynamic reconfiguration* of the FCR for each A_i, $i = 1, 2,\ldots, m$ is implemented according to requested mode.

In this consideration, we assume that all modes of operation are known ahead of time and certain FCR configurations are designed for each mode—$M_1, M_2,\ldots, M_m$. Each mode performs subset of task functions, $\{F\}_j$, associated with this mode. In other words, $\{F\}_j \rightarrow \{TA_j \,\&\, DS_j\}$.

Thus, each task determines a certain set of functions $\{F\}$, which consists of all functions included in subsets of all modes: $\{F\}_j$, where $j = 1, 2,\ldots, m$. It is understood that only one of the aforementioned modes can be active at a time. And each mode is active during the period of time, T_j, $j = 1, 2,\ldots, m$. This period of time may be determined or unknown ahead of time but during this period the configuration of the FCR is static. Thus, each mode of operation M_j is always associated with (1) a variant of the algorithm and data structure $\{F\}_j$ and (2) the set of constraints $\{SC\}_j$:

$$M_j \rightarrow \{F\}_j \text{ and } \{SC\}_j, \quad j = 1, 2,\ldots, m.$$

The configuration of the FCR for each mode of operation assumes configuration of the architecture A_j optimized for the characteristics and performance constraints of mode M_j. This architecture A_j utilizes the determined number of elements of each type of resources R_i in mode M_j: UR_{ij}, $i = 1, 2,\ldots, n$; $j = 1, 2,\ldots, m$. These amounts of resources are dedicated for the period of mode M_j operation—T_j. Hence, the FCR architecture, A_j, optimized for the mode M_j employs a certain set of FCR elements $\{UR_{ij}\}$ statically configured for the period T_j.

2.4.3 Efficiency of Static and Dynamic Reconfiguration of the FCR

In case of static reconfiguration, the entire set of resources in FCR must be stopped to allow safe reconfiguration of the FCR. Hence, reconfiguration time causes hardware overhead due to stalled resources in FCR for all reconfiguration period. Alternatively, the dynamic reconfiguration approach can be implemented to reduce the aforementioned hardware overhead in multimodal applications (tasks). However, dynamic reconfiguration cannot exclude the hardware overhead associated with reconfiguration of resources but makes it possible to perform run-time reconfiguration. To avoid interruption of the task execution, no element of UR_{ij} can be suspended during T_j. Therefore, when a request for the next mode of operation (e.g., M_k) occurs, the reconfiguration process should avoid elements of the $\{UR_{ij}\}$ set currently being used. Obviously, additional elements of each resource type R_i should be reserved, to accommodate the reconfiguration from architecture A_i to A_k, referred to as ΔUR_i. Thus, the maximum amount of additional R_i elements— $\max\{\Delta UR_{ik}\}$, $i = 1, 2,\ldots, n$ and $k = 1, 2,\ldots, m$—can be considered as the hardware overhead in the FCR. This *hardware overhead is the cost for noninterrupted mode switching* in RTR RCS. In this case, mode switching process is performed by reconfiguration of additional resources at run-time. The general picture of the aforementioned mode switching process is depicted in Figure 2.5.

After the initial configuration period T_{conf_init}, the task execution runs noninterrupted. At certain moments in time, a mode switch is requested. The mode switch process begins with the reconfiguration of the reserved resources in some part(s) of FCR according to the upcoming (requested) mode of task. The reconfiguration of the FCR to mode M_j requires a period of time T_{conf_j} associated with the configuration of the needed resources. Figure 2.5 shows two reconfiguration periods: T_{conf_j} (necessary to complete resource reconfiguration for switching to mode M_j) and T_{conf_k} (necessary to complete resource reconfiguration for switching to mode M_k). These reconfiguration periods are *shadowed* (hidden) by task execution periods. Task execution in initial mode (T_{init}) *shadows* the configuration period T_{conf_j} while task execution in mode $M_j - T_j$ hides the configuration period T_{conf_k}. However, in such cases the reconfiguration time for the requested mode (e.g., T_{conf_j}) should be less than the mode execution time. For example, for

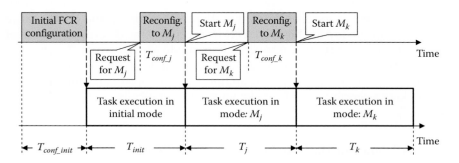

FIGURE 2.5
Timing diagram of configuration, task execution, and mode switching processes.

video-processing applications, the mode reconfiguration time should be less than or equal to the video-frame execution period. In network processing applications, reconfiguration time should be less than or equal to the packet processing time, etc.

If this condition can be satisfied, the RUF for the DPR FCR—RUF_{dpr}—can be calculated by the following equation:

$$RUF_{dpr} = \frac{\sum_{j=1}^{m}\left[\sum_{i=1}^{n}(UR_{ij} \times a_i) \times T_j\right]}{\sum_{i=1}^{n}(NR_i \times a_i) \times \left(\sum_{j=1}^{m}T_j + T_{conf\_init}\right)} \times 100\% \qquad (2.5)$$

In this equation, T_{conf_init} is the configuration time for the initial FCR configuration. However, all periods of further FCR reconfiguration (T_{conf}) are excluded from the equation as a result of the aforementioned condition: $T_{conf\_k} \leq T_j$ for $j, k = 1, 2,\ldots, m$.

Thus, the nominator of RUF_{dpr} represents the sum of all areas of resources used within all modes of operation. The denominator of this equation consist of the total areas of all resource elements $\sum_{i=1}^{n}(NR_i \times a_i)$ multiplied by the task execution time in all modes of operation, $\sum_{j=1}^{m}T_j$, plus the initial configuration time of the FCR, T_{conf_init}.

It is interesting to see that the sum of the execution time for all modes of operation is equal to the complete task execution time, T_{exe}, for a statically configured FCR: $T_{exe} = \sum_{j=1}^{m}T_j$ for $j = 1, 2,\ldots, m$. Therefore, Equation 2.5 can be presented as

$$RUF_{dpr} = \frac{\sum_{j=1}^{m}\left[\sum_{i=1}^{n}(UR_{ij} \times a_i) \times T_j\right]}{\sum_{i=1}^{n}(NR_i \times a_i) \times (T_{exe} + T_{conf\_init})} \times 100\% \qquad (2.6)$$

Here, the denominator is the same as in Equation 2.4, defined for statically reconfigurable RCS.

To clarify the aforementioned RUF_{dpr} computation, let us consider an example of an RTR multimodal RCS. In this example, the same FCR organization that was presented in the example for RUF_{SRCS} in Section 2.3.3 will be assumed. For this example, let us consider a video-stream processing application that can work in four different modes. Data-stream applications usually require nonstop stream processing. However, the set of functions being executed on the stream may be different and depend on the mode of operation. At the same time, the mode switching period should not exceed the period of one video-frame. During the mode switching period, the video-stream should be captured and processed without interrupts. As well, the output video-stream should also be generated without any pause. After the aforementioned constraints, let us consider the following functional and technical specifications:

1. The application is associated with mobile robotic vision and control. The application consists of four task modes:
   a. M_1 is the initial mode in this application and performs scene observation and display.
   b. M_2 is the next mode and performs detection of objects with a certain shape in the scene.
   c. M_3 follows M_2 and determines the target object to be approached.
   d. M_4 conducts tracking of the selected object while the system approaches said object.

   This application is quite common for mobile robotic systems; hence, it is used as an example.

2. Table 2.1 presents a specification of resource utilization for each type of resource and mode combination. As well, it provides periods of task execution in each mode.*

3. Table 2.2 provides a list of functions associated with each task mode (e.g., F_1 [video-frame acquisition], F_2 [edge detection of the object], F_3 [centroid calculation of the object]).*

4. Table 2.3 describes resource utilization for each function: F_1, F_2, F_3, F_4, F_5, and F_6.*

According to the aforementioned specifications, it is possible to calculate the RUF for RCS as follows:

TABLE 2.1

Resource Utilization and Active Time of Each Application Mode of Operation

Application Mode #	Resources Used for Application Mode			Task Mode Active Time (Max) T_j, for $j = 1, 2, 3, 4$
	CLBs	MACs	BRAMs	
Mode #1—M_1	656	58	6	3333 ms (100 video-frames)
Mode #2—M_2	590	82	8	333 ms (10 video-frames)
Mode #3—M_3	750	76	8	133 ms (4 video-frames)
Mode #4—M_4	878	84	8	Up to 10 s (300 video-frames)

---

* *Note 1*: All quantitative specifications presented in Tables 2.1 through 2.3 are not taken from any actual application. They are arbitrarily selected for the purpose of illustrating the RUF_{dpr} calculation process for this particular example.

TABLE 2.2

List of Functions associated with Each Mode of Operation

| Application Mode # | Functions Involved in Application Mode | | | | | |
	F_1	F_2	F_3	F_4	F_5	F_6
Mode #1—M_1	Yes	No	No	No	Yes	Yes
Mode #2—M_2	Yes	Yes	Yes	No	No	No
Mode #3—M_3	Yes	Yes	No	Yes	No	No
Mode #4—M_4	Yes	Yes	No	Yes	Yes	No

TABLE 2.3

Resource Utilization for Each Function of the Application

| Functions of the Application | Resources Required for Function Implementation | | | Configuration Time for Function (ms) |
	CLBs	MACs	BRAMs	
F_1	264	42	4	10.84
F_2	108	16	2	4.68
F_3	218	24	2	6.58
F_4	378	18	2	7.58
F_5	128	8	0	2.08
F_6	264	8	2	5.44

1. Resource utilization for each mode and total resources utilization during task execution:

   a. For the initial mode M_1

$$\sum_{i=1}^{n} (UR_{i1} \times a_i) \times T_1 = (\text{Number of CLB used in } M_1 \times 10 \text{ au}$$

$$+ \text{Number of MACs used in } M_1 \times 100 \text{ au}$$
$$+ \text{Number of BRAMs} \times 1000 \text{ au})$$
$$\times \text{Active time of the mode } M_1 \ (T_1)$$
$$= (656 \text{ CLBs} \times 10 \text{ au} + 58 \text{ MACs} \times 100 \text{ au}$$
$$+ 6 \text{ BRAM} \times 1000 \text{ au})$$
$$\times 3.333 \text{ s} = 61,193.88 \text{ au-s}$$

The same calculations for modes M_2, M_3, and M_4 will result in the following:

   b. For mode M_2

$$\sum_{i=1}^{n} (UR_{i2} \times a_i) \times T_2 = (590 \times 10 + 82 \times 100 + 8 \times 1000) \times 0.333 = 7359.3 \text{ au-s}$$

c. For mode M_3

$$\sum_{i=1}^{n}(UR_{i3}\times a_i)\times T_3 = (750\times10+76\times100+8\times1000)\times0.133 = 3072.3 \text{ au-s}$$

d. For mode M_4

$$\sum_{i=1}^{n}(UR_{i4}\times a_i)\times T_4 = (878\times10+84\times100+8\times1,000)\times10\,\text{s} = 251,800 \text{ au-s}$$

Thus, the nominator in Equation 2.6 is equal to

$$\sum_{j=1}^{m}\left[\sum_{i=1}^{n}(UR_{ij}\times a_i)\times T_j\right] = 61,193.88 \text{ au-s} + 7,359.3 \text{ au-s}$$
$$+ 3,072.3 \text{ au-s} + 251,800 \text{ au-s} = 323,425.48 \text{ au-s}$$

2. The task execution time is the sum of periods of all four modes of operation:

$$T_{exe} = \sum_{j=1}^{m}T_j = 3.333 \text{ s} + 0.333 \text{ s} + 0.133 \text{ s} + 10 \text{ s} = 13.799 \text{ s}$$

The initial configuration time is usually equal to the start-up time of the entire FCR and depends on the volume of the configuration bit-file V_{conf_j} and the bandwidth of the configuration bus—BW_{conf}: $T_{conf\_j} = V_{conf\_j}/BW_{conf}$.

For this example, let us assume that the configuration bandwidth allows configuration of 1,000,000 area units (au) per second. The entire FCR consisting of 1000 CLBs, 100 MACs, and 10 BRAMs can be initially configured within

$$T_{conf\_init} = \sum_{i=1}^{n}(NR_i\times a_i)/BW_{conf} =$$
$$= (1,000 \text{ CLB}\times10 \text{ au} + 100 \text{ MAC}\times100 \text{ au}$$
$$+ 10 \text{ BRAM}\times1,000 \text{ au})/1,000,000 \text{ au/s} = 0.03 \text{ s} *$$

3. Therefore, according to Equation 2.6, the RUF for the RTR FCR considered in this example can be obtained by the following equation:

---

* *Note 2*: The initial configuration time calculation is simplified for this example. The detailed description of configuration schemes and configuration time will be given in Chapter 8.

$$RUF_{dpr} = \frac{\sum_{j=1}^{m}\left[\sum_{i=1}^{n}(UR_{ij}\times a_i)\times T_j\right]}{\sum_{i=1}^{n}(NR_i\times a_i)\times(T_{exe}+T_{conf\_init})}\times 100\%$$

$$= \frac{323,425.48\ au\_s\times 100\%}{30,000\ au\times(13.799+0.3)s} = 76.47\%$$

Let us compare the DPR FCR presented here with the SFCR described in Section 2.3.3. In this case, all functions listed in Table 2.3 need to be placed (as components) in the FCR. Mode switching can be provided by a multiplexing scheme. In this case, the total amount of resources accommodating the entire set of functions $\{F_1, F_2, F_3, F_4, F_5, F_6\}$ is equal to the sum of required CLBs plus the sum of all MACs and BRAMs associated with the aforementioned functions: 1360 CLBs, 116 MACs, and 12 BRAMs.

It is interesting to see that the total amount of each type of resource exceeds the initial amount considered for the DPR FCR (1000 CLBs, 100 MACs, and 10 BRAMs), even without considering the added overhead of the multiplexers needed by the SFCR.

In this case, the total amount of resources in the according to the denominator of Equation 2.3 is

$$\sum_{i=1}^{n}(NR_i\times a_i)=1,360\ CLB\times 10\ au+116\ MACs\times 100\ au$$

$$+12\ BRAMs\times 1,000\ au = 37,200\ au$$

Thus, the comparison of RUFs for the aforementioned SFCR and DPR FCR (according to Table 2.1) for each mode is as follows:

1. For mode M_1: RUF_{srcs} (M_1) $= \dfrac{\begin{array}{c}656\times 10\ au+58\times 100\ au\\+6\times 1,000\ au\end{array}}{37,200\ au}\times 100\% = 49.35\%$

   in comparison to RUF_{dpr} (M_1) $= \dfrac{\begin{array}{c}656\times 10\ au+58\times 100\ au\\+6\times 1,000\ au\end{array}}{30,000\ au}\times 100\% = 61.2\%$

2. For mode M_2: RUF_{srcs} (M_2) $= \dfrac{\begin{array}{c}590\times 10\ au+82\times 100\ au\\+8\times 1,000\ au\end{array}}{37,200\ au}\times 100\% = 59.41\%$

   in comparison to RUF_{dpr} (M_2) $= \dfrac{\begin{array}{c}590\times 10\ au+82\times 100\ au\\+8\times 1,000\ au\end{array}}{30,000\ au}\times 100\% = 73.7\%$

3. For mode M_3: $\mathrm{RUF_{srcs}} = \dfrac{\begin{array}{c}750 \times 10\ \mathrm{au} + 76 \times 100\ \mathrm{au} \\ + 8 \times 1{,}000\ \mathrm{au}\end{array}}{37{,}200\ \mathrm{au}} \times 100\% = 62.1\%$
(M_3)

in comparison to $\mathrm{RUF_{dpr}} = \dfrac{\begin{array}{c}750 \times 10\ \mathrm{au} + 76 \times 100\ \mathrm{au} \\ + 8 \times 1{,}000\ \mathrm{au}\end{array}}{30{,}000\ \mathrm{au}} \times 100\% = 77\%$
(M_3)

4. For mode M_4: $\mathrm{RUF_{srcs}} = \dfrac{\begin{array}{c}878 \times 10\ \mathrm{au} + 84 \times 100\ \mathrm{au} \\ + 8 \times 1{,}000\ \mathrm{au}\end{array}}{37{,}200\ \mathrm{au}} \times 100\% = 67.69\%$
(M_4)

in comparison to $\mathrm{RUF_{dpr}} = \dfrac{\begin{array}{c}878 \times 10\ \mathrm{au} + 84 \times 100\ \mathrm{au} \\ + 8 \times 1{,}000\ \mathrm{au}\end{array}}{30{,}000\ \mathrm{au}} \times 100\% = 83.9\%$
(M_4)

Based on these calculations, it is interesting to see that the SFCR resource utilization in all modes of operation is less than that of the DPR FCR. At the same time, the total amount of required resources is 24% higher when compared to the RTR FCR—37,200 au (in static FCR) versus 30,000 (in DPR FCR). The aforementioned resource and *RUF* calculations do not include the *additional hardware resources needed for mode switching multiplexors and the associated control circuits* required by the SFCR. Therefore, the real amount of hardware resources in the static FCR will be even higher when compared with the DPR FCR.

The reason for these effects is quite clear—the SFCR must include all hardware circuits for all possible modes of operations. Therefore, all functional components must be deployed in the initial configuration of the FCR. However, not all of them are used for all modes of operation. Hence, a portion of functional components are not in active use. In the example earlier, during operating mode M_1, functions F_1, F_5, and F_6 are the only active functions; all other functions (and, thus, their associated resources) are not in use. The unused amount of resources depends on the mode of operation and can be thought of as a *dynamic hardware overhead* in a multimodal SFCR.

The more modes of operations are accommodated by the RCS, the higher this dynamic overhead is. At the same time, a dynamic overhead may exist in DPR FCR as well. However, the cause of this overhead is different than that found in SFCR. According to Figure 2.5, there are two parallel processes running in FCR:

1. The task execution employing resources required for the current mode of operation

2. The reconfiguration process that configures reserved resources for the upcoming mode

The reserved resources being configured are not in use in the current mode of operation—M_j. Hence, they should be considered as a hardware overhead—ΔUR_{ik}—for each R_i, $i = 1, 2,\ldots,n$, before switching to the next mode M_k. Obviously, this hardware overhead depends on the current and upcoming modes of operation and is, therefore, dynamic in nature.

Let us now consider this aspect when computing the *RUF* for the DPR FCR presented earlier.

Based on Table 2.2, mode M_1 activates a subset of functions including F_1, F_5, and F_6. Table 2.1 shows that 656 CLBs (out of 1000), 58 MACs (out of 100), and 6 BRAMs (out of 10) have been used. To switch to mode M_2, functions F_2 and F_3 must first be configured into the FCR. These components will need (1) $108 + 218 = 326$ CLBs, (2) $16 + 24 = 40$ MACs, and (3) $2 + 2 = 4$ BRAMs. In other words, the following numbers of additional resources are needed:

1. R_1 (CLB) for switching to the M_2 $-\Delta UR_{1,2} = 326$
2. R_2 (MAC) for switching to the $M_2-\Delta UR_{2,2} = 40$
3. R_3 (BRAM) for switching to the $M_2-\Delta UR_{3,2} = 4$

It must be mentioned that the condition for switching from mode M_j to mode M_k (according to Figure 2.5) is that the sum of resources in use (for current mode M_j) and resources under reconfiguration for upcoming mode M_k should be less than or equal to the total available resources in the FCR. In the example earlier, this condition is satisfied as shown:

1. 656 CLBs (used in mode M_1) + 326 CLBs (in reconfiguration to M_2) = $982 < 1000$ CLB.
2. 58 MACs (used in mode M_1) + 40 MACs (in reconfiguration to M_2) = $98 < 100$ MACs.
3. 6 BRAMs (used in mode M_1) + 4 BRAMs (in reconfiguration to M_2) = $10 = 10$ BRAMs.

The same calculations should be done for all mode switching processes. For example, switching from mode M_2 to M_3 requires the configuration of resources for function F_4 while F_1, F_2, and F_3 are active. Since M_2 employs 590 CLBs, 82 MACs, and 8 BRAMs, there are enough resources in the FCR to accommodate configuration of F_4, which requires 378 CLBs, 18 MACs, and 2 BRAMs. Thus, the maximum number of elements of each type of resources R_i, $i = 1, 2,\ldots,n$, in the FCR should be determined as the sum of max$\{UR_{ij}\}$ + max$\{\Delta UR_{i,k}\}$ for all resources and modes $(M_1, M_2,\ldots,M_m)$. Thus, max$\{\Delta UR_{ik}\}$ will be the maximum hardware overhead for resource type R_i in this FCR.

Summarizing the aforementioned comparative analysis of an SFCR versus a dynamically reconfigurable (DPR) FCR, the following can be concluded:

1. SFCR and DPR FCR organizations can provide almost equal mode switching times. However, the response time of the DPR FCR is much higher compared with the SFCR due to the relatively long reconfiguration time of the new functional components needed for the upcoming mode of operation. Furthermore, the reconfiguration time in the DPR FCR depends on the set of functions used by the current and upcoming modes. This time may vary within a certain range.

2. Utilization of resources in the DPR FCR is better than in the SFCR. The RUF increases together with the number of modes and functions that the system performs. For multimodal applications, the amount of resources required for a DPR FCR could be much less than in an SFCR designed for the same application.

3. Both static and DPR FCR have a *dynamic resource overhead* associated with their operation. However, the reasons for the appearance of this overhead in the static and the DPR FCR are different. If only one set of functions should be executed on the FCR, reconfiguration is not needed at all. In the case where multiple sets of functions must be executed on an SFCR, all functionalities must be implemented in the form of hardware circuits. Thus, the current mode of operation activates the associated set of functions while the rest of functions and associated circuits are not in operation. Therefore, the more modes of operation the static FCR should support, the higher hardware overhead will be due to inactive functional components. On the other hand, the overhead in a DPR FCR depends only on the additional functions to be configured for the upcoming mode. The rest of the functionalities are stored in the form of configuration files (bit-streams) in memory. Thus, the overhead in a DPR FCR does not grow in proportion to the number of modes and associated functions deployed in the FCR.

4. Due to the aforementioned reasons, the total hardware cost for multimodal SFCR is always higher than for the respective DPR FCR. This cost may grow proportionally to the number of modes in SFCR, which is not a case for DPR FCR. In all cases, cost-efficiency of the DPR FCR is higher than SFCR.

2.5 Spatial versus Temporal Partitioning of Resources in an FCR

The *dynamic overhead* of resources in an FCR discussed in the previous section is the implicit *penalty* for the ability to perform function-switching or mode-switching operations in the RCS without interruption of processing tasks. In an SFCR, these functional components are placed in *space* within the FCR or in the *spatial domain*. If functional components are configured in the FCR *only* for the period of the associated mode of operation, these components are allocated in the *temporal domain*.

Therefore, the concept of a *space–time continuum* can be applied to reconfigurable systems. Indeed, the flexibility of the resource configuration process associated with an FCR allows

1. Execution of the entire task in FCR at the highest possible speed
2. Execution of the task part by part, while spending fewer resources but more time

This latter approach assumes proper partitioning of the task into segments, as well as the associated partitioning of resources in the FCR in space and time. To facilitate further discussion, a number of definitions are provided, dealing with task and resource partitioning.

Definition 2.13 *A segment of a task* is a part of a task that consists of a determined function or set of functions and an associated data structure.

Thus, a segment of a task is considered as a subtask of the entire task. Each segment is defined through (1) a set of functions, (2) a set of initiation and termination conditions, and (3) an associated data structure.

Definition 2.14 *Spatial partitioning* of configurable resources represents the mapping task segments on different areas of the FCR in the *spatial domain*.

In other words, task segments are allocated to dedicated areas (partitions) of hardware resources according to the resource requirements of each segment.

Definition 2.15 *Temporal partitioning* of configurable resources represents the process of allocating different task segments *to the same set of resources* in the *temporal domain*.

In other words, temporal partitioning of resources assumes that task segments are placed on the FCR in a certain sequence and for required periods of time.

To illustrate the aforementioned concepts of spatial and temporal partitioning of resources in the FCR, let us return to the example of multimodal heterogeneous RCS discussed in Section 2.4.3.

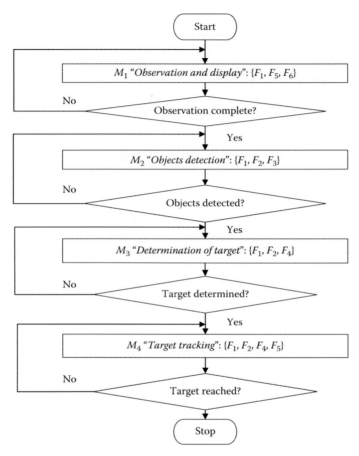

FIGURE 2.6
Flowchart of example application.

Table 2.2 presents the complete set of functions needed by the task and their distribution in each task mode. The structure of this task algorithm is shown in Figure 2.6 in the form of a flow chart.

The application (task) always starts from mode M_1, *observation and display*. This mode requires the processing of video-streams using functions F_1, F_5, and F_6. After the accumulation of a predetermined number of video-frames, the system generates the request for object(s) detection. When the request signal is detected, the system should switch to mode M_2 where a different set of functions is active—F_1, F_2, and F_3.

When all the objects with certain shapes are detected, the system can start mode M_3, *determination of target object*. In this mode, functions F_1 and F_2 remain the same but function F_3 is replaced by F_4.

Finally, when the target is selected, the system starts mode M_4, which allows it to approach the target object while at the same time tracking the target position. During this mode, F_5 is added to the set of active functions. When the target object is reached, the application is complete.

Let us consider three possible implementations of the aforementioned application:

1. SFCR with spatial partitioning of resources into components
2. DPR (run-time) FCR with spatial and temporal partitioning of resources into functional components
3. Implementation in the form of a non-RTR FCR with temporal partitioning of resources into task segments

The first two implementations have been discussed in Section 2.4 in an attempt to figure out pros and cons of static versus dynamic FCR configuration. Both these implementations were considered as run-time mode switching systems. It was assumed that the mode switching time should be close to the time of circuit multiplexing (in the range of nanoseconds). However, if the application specification does not have such constraints, the third implementation, using temporal partitioning of the FCR, could be an effective solution. Let us consider each of the aforementioned implementations and analyze their benefits and drawbacks.

2.5.1 Statically Configurable FCR with Spatial Partitioning of Resources

This type of system implementation is the most common method of deployment of applications, which require real-time high-performance dedicated processing circuits. The usual platforms for such system implementation in the on-chip level of FCR are statically configurable FPGA devices. The configuration memory of these FPGA devices is based on SRAM. There are many examples of such FPGAs produced by Xilinx Inc., Altera, Lattice Semiconductor, etc. Most of FPGAs have heterogeneous architecture where FCR consist of CLBs, MACs, BRAMs, and other function-specific units as was described in previous sections. Figure 2.7 shows one of possible area distributions of each type of resources allocated to functional components performing task functions (according to Table 2.3). In this figure, the system implementation is shown in mode M_1 where components associated with task functions F_1, F_5, and F_6 (marked in dark grey color) are in operation. It is easy to see that more than half of available resources are not activated. The RUF for this mode was calculated in Section 2.4.3 as 49.35%.

The mode switching process is based on multiplexing and demultiplexing circuits (not shown in Figure 2.7). Therefore, mode switching can be done in the nanosecond time range.

Statically configured (at start-up time) FCR

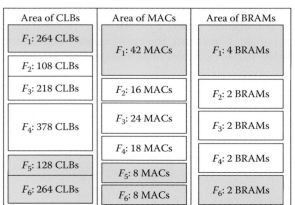

FIGURE 2.7
Spatial allocation of functional components in the statically configured field of configurable resources.

It is necessary to mention that in the FPGA platform, all nonactive components are still consuming power. The larger the FPGA device used for spatially allocated components, the higher the cost of the system and associated power consumption. As well, large FPGAs have longer routing lines between components and more routing switches or routing multiplexors when compared with smaller devices. Therefore, large FPGAs result in higher propagation delays in communication lines when compared to smaller FPGA devices of the same family. Hence, in general, circuits deployed in large FPGAs operate in lower frequencies than those being deployed in smaller FPGA devices.

2.5.2 Run-Time Reconfigurable FCR with Spatial and Temporal Partitioning of Resources

A system implementation based on the run-time reconfiguration of a DPR FCR in response to a requested mode is often called dynamic reconfiguration of resources, and an RCS based on this concept is called a DRCSs. The hardware platform usually used for this class of RCS is a PR FPGA. The heterogeneous organization of the FCR in this type of FPGA is a common trait for most families of such devices. In general, the FCR is divided into smaller portions often called *frames*. Each frame can be configured individually and on the fly. It means that the architecture of a PR FPGA allows reconfiguration of any number of resource frames while other resources continue operation. This aspect will be discussed in detail in Chapter 7. Figure 2.8 depicts spatial allocation of all functional components associated with functions F_1, F_5, and F_6, which are involved in data-processing

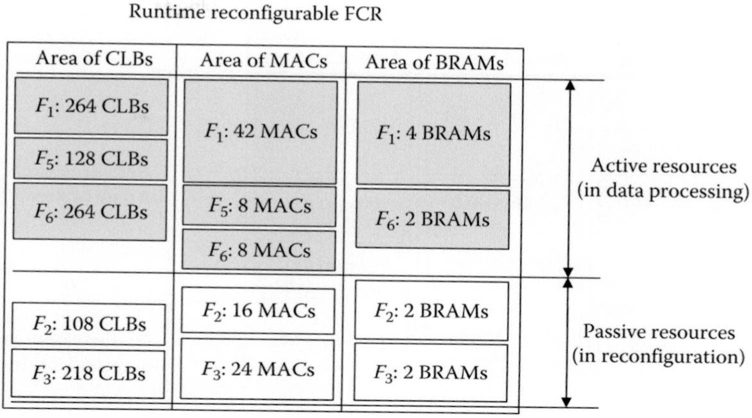

Runtime reconfigurable FCR

Area of CLBs	Area of MACs	Area of BRAMs	
F_1: 264 CLBs	F_1: 42 MACs	F_1: 4 BRAMs	Active resources (in data processing)
F_5: 128 CLBs			
F_6: 264 CLBs	F_5: 8 MACs	F_6: 2 BRAMs	
	F_6: 8 MACs		
F_2: 108 CLBs	F_2: 16 MACs	F_2: 2 BRAMs	Passive resources (in reconfiguration)
F_3: 218 CLBs	F_3: 24 MACs	F_3: 2 BRAMs	

FIGURE 2.8
Spatial allocation of functional components in a dynamic partially reconfigurable field of configurable resources for mode M_1.

in mode M_1. While the system performs mode M_1, additional reserved resources (passive during mode M_1) can be configured with functional components for functions F_2 and F_3, which are required for the upcoming mode M_2.

When configuration of components F_2 and F_3 is complete, the system can switch to mode M_2 once mode M_1 is complete (according to the algorithm shown in Figure 2.6). After initiation of mode M_2, resources associated with functions F_5 and F_6 can be released, as these functions are not active in the current mode (M_2). These resources can be considered as reserved and are available for use by upcoming modes of operation, specifically mode M_3:

1. In gray: active components involved in data-processing in mode M_1
2. In white: areas of reserved resources that are currently being configured with the functional components required for mode M_2

This mode utilizes components performing functions F_1, F_2, and F_4 (according to Table 2.2). Functional components F_1 and F_2 are already configured and are working in the current mode (M_2). Hence, only component for function F_4 should be configured into the FCR prior to switching to mode M_3. The spatial allocation of components in the FCR at this stage of system operation is shown in Figure 2.9.

1. In gray: active components involved in data-processing in mode M_2
2. In white: areas of reserved resources that are currently being configured with the functional components required for mode M_3

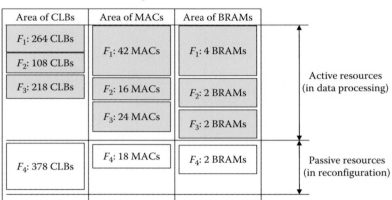

Runtime reconfigurable FCR

Area of CLBs	Area of MACs	Area of BRAMs	
F_1: 264 CLBs	F_1: 42 MACs	F_1: 4 BRAMs	Active resources (in data processing)
F_2: 108 CLBs			
F_3: 218 CLBs	F_2: 16 MACs	F_2: 2 BRAMs	
	F_3: 24 MACs	F_3: 2 BRAMs	
F_4: 378 CLBs	F_4: 18 MACs	F_4: 2 BRAMs	Passive resources (in reconfiguration)

FIGURE 2.9
Spatial allocation of functional components in a dynamic partially reconfigurable field of configurable resources for mode M_2.

Summarizing the considered examples earlier, it is possible to see the following:

1. There is *no static architecture* spatially deployed in the FCR. The architecture dynamically changes at run-time and provides noninterrupting switching to the requested mode of operation. In effect, some components appear and some disappear at certain moments of time inside the FCR. Thus, the spatial distribution of resources in a DPR FCR must always incorporate a temporal element.

2. The operation of the RTR FCR consists of *two parallel processes*: the data-execution process and the reconfiguration process of components for the next mode.

3. The amount of resources needed for task implementation in a DPR FCR *is less* than in an SFCR because FCR should include the resources to deploy functions for the current mode of operation and additional resources for difference of functions required for the next mode(s). Certainly, this amount of required resources will be less in the case when all functions for all modes must be deployed in the FCR upfront. As the calculations for this example have shown earlier, the ratio of resource requirements is equal to 37,200/30,000 = 1.24 or 24% less. This is because all nonused (in the current mode) components are represented in the form of configuration bit-files—a *virtual form*—as opposed to the actual physical hardware.

4. To provide a noninterrupting mode switching operation, reconfiguration of reserved resources should start ahead of time. Hence, the next mode of operation should be *predicted* to insure that a sufficient

time window is available prior to the upcoming mode switch, where configuration activities can take place (as shown in Figure 2.5). In this context, the *mode prediction time* is the period of time between the moment when the next mode of operation is determined and the time when the task in the current mode completes execution. If both conditions can be satisfied, implementation of the task in a DPR FCR will be more effective when compared to a statically configured FCR.

2.5.3 Non-Run-Time Reconfigurable FCR with Temporal Partitioning of Resources

Let us consider the mode prediction aspect in detail. According to the discussion earlier , two conditions should be satisfied to ensure a noninterrupting mode switching process:

1. Task execution time in mode $M_j - T_j$ should be greater than the reconfiguration time for the next mode of operation $M_k - T_{conf\_k}$: $T_j > T_{conf\_k}$ for $j, k = 1, 2,..., m$
2. Mode prediction time for mode $M_k - T_{pred\_k}$ should be greater than the reconfiguration time for this mode of operation $M_k - T_{conf\_k}$: $T_{pred\_k} > T_{conf\_k}$ for $j, k = 1, 2,..., m$

If any of these conditions are violated, a delay will occur between the end of mode M_j and initiation of the next mode M_k. Figure 2.10 illustrates how violation of the second condition causes the delay in the mode switching and, thus, a mode M_k waiting time—T_{wait_k}. The aforementioned process is

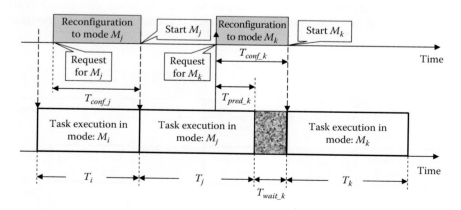

FIGURE 2.10
Delay in task mode switching caused by violation of the second condition for noninterrupted task execution process.

similar to the process shown in Figure 2.5. The only difference is that the reconfiguration time shown in Figure 2.10 is greater than the mode prediction time.

If the application does not require noninterrupting mode switching and allows a mode waiting time equal to or higher than the mode reconfiguration time, temporal partitioning of resources into the functional components associated with a given task mode can be considered as a solution. In this case, only those functional components associated with the current mode are allocated in the FCR. In contrast to the DPR FCR method of operation, the process of task execution in the current mode and the process of resource reconfiguration for the next mode are *running sequentially* as shown in Figure 2.11.

This approach does not require mode prediction or a PR FCR organization. The mode switching process looks like a sequence of SFCRs as shown in Figure 2.10. Therefore, this approach can be considered as *statically reconfigurable* FCR. In contrast with *SFCR* where configuration of resources is done only at the start-up time, statically reconfigurable organization of the FCR assumes multiple configurations and reconfigurations of resources in FCR according to respective modes of operation. This scheme is called multibooting and will be discussed in Chapter 8.

Same as in previous DPR FCR, each mode of operation consists of two periods:

1. The FCR configuration period T_{conf_j}, $j = 1, 2, ..., m$.
2. The task mode execution period, T_j, when the FCR is spatially partitioned to incorporate the functions needed by task mode M_j.

Another advantage of the non-RTR FCR with temporally partitioned resources is a minimization of resource utilization when compared with dynamically or SFCR organizations. Indeed, if reconfiguration for the next

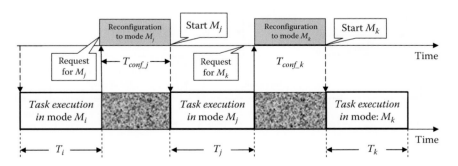

FIGURE 2.11
Sequential processes of task mode execution and reconfiguration for the next mode.

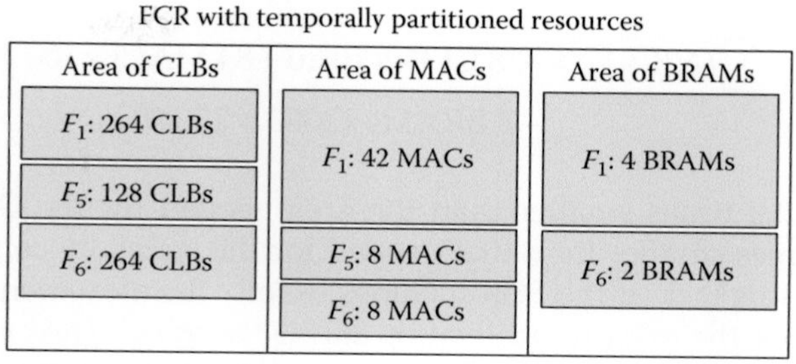

FCR with temporally partitioned resources

Area of CLBs	Area of MACs	Area of BRAMs
F_1: 264 CLBs	F_1: 42 MACs	F_1: 4 BRAMs
F_5: 128 CLBs		
F_6: 264 CLBs	F_5: 8 MACs	F_6: 2 BRAMs
	F_6: 8 MACs	

FIGURE 2.12
Allocation of functional component in the field of configurable resources in mode M_1.

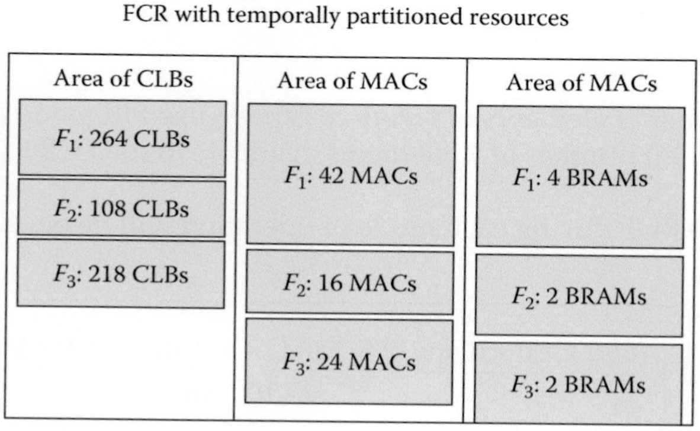

FCR with temporally partitioned resources

Area of CLBs	Area of MACs	Area of MACs
F_1: 264 CLBs	F_1: 42 MACs	F_1: 4 BRAMs
F_2: 108 CLBs		
F_3: 218 CLBs	F_2: 16 MACs	F_2: 2 BRAMs
	F_3: 24 MACs	F_3: 2 BRAMs

FIGURE 2.13
Allocation of functional component in the field of configurable resources in mode M_2.

mode could be started after completion of current mode execution, there is no need to have any reserved resources for run-time reconfiguration. In turn, the total amount of resources in the FCR is smaller when compared to other types of FCR organizations. For each type of resource, the FCR should provide the maximum number of R_i units used in any task mode (M_j: $NR_i = \max\{UR_{ij}\}$), for any type of resources (R_i, $i = 1, 2,\ldots,n$), and any mode M_j, $j = 1, 2,\ldots,m$.

For the example discussed in Section 2.4 and according to Tables 2.2 and 2.3, the spatial allocation of components for modes M_1 and M_2 is shown in Figures 2.12 and 2.13.

The maximum amount of resources that needs to be included in the FCR can be determined according to Table 2.1. Of all modes, Mode 4 requires the maximum of each type of resource, that is, 878 CLBs, 84 MACs, and 8 BRAMs.

Hence, the FCR should not have more than these amounts of these types of resources. Thus, the total area of this FCR is equal to

$$\sum_{1}^{n}(NR_i \times a_i) = 878 \text{ CLBs} \times 10 \text{ au} + 84 \text{ MACs} \times 100$$
$$+ 8 \text{ BRAMs} \times 1000 = 25,180 \text{ au}$$

This area is 1.2 times smaller than the area needed for the DPR FCR and almost 1.5 times smaller than that needed for the statically configured FCR required for this task. If the FCR includes exactly the aforementioned calculated resources, the resource utilization *during the period of task mode execution* according to Table 2.1 for each mode can be calculated as follows:

$$\text{RUF}_{exe}(M_j) = \frac{\sum_{1}^{n}(UR_i \times a_i)j}{\sum_{1}^{n}(NR_i \times a_i)} \times 100\%, \tag{2.7}$$

where
  UR_i is number of elements of resource type R_i used in mode M_j execution
  NR_i is the total number of R_i elements available in the FCR

Therefore the RUF during each mode of operation will be equal to

For M_1:

$$\text{RUF}_{exe}(M_1) = \frac{(656 \text{ CLB} \times 10 \text{ au} + 58 \text{ MAC} \times 100 \text{ au} + 6 \text{ BRAM} \times 1000 \text{ au})}{25,180 \text{ au}}$$
$$\times 100\% = 72.9\%$$

For M_2:

$$\text{RUF}_{exe}(M_2) = \frac{(590 \text{ CLB} \times 10 \text{ au} + 82 \text{ MAC} \times 100 \text{ au} + 8 \text{ BRAM} \times 1000 \text{ au})}{25,180 \text{ au}}$$
$$\times 100\% = 87.8\%$$

For M_3:

$$\text{RUF}_{exe}(M_3) = \frac{(750 \text{ CLB} \times 10 \text{ au} + 76 \text{ MAC} \times 100 \text{ au} + 8 \text{ BRAM} \times 1000 \text{ au})}{25,180 \text{ au}}$$
$$\times 100\% = 91.7\%$$

For M_4: $\text{RUF}_{exe}(M_4) = 100\%$

According to these calculations, utilization of resources is better than in the SFCR or the DPR FCR. However, this is true only for the period of task execution. If the reconfiguration time for the mode of operation is taken into account according to Equation 2.4, the value of the *RUF* will be lower for all modes. Indeed, the RUF for each mode can be calculated by using an equation similar to Equation 2.4:

$$\text{RUF}(M_j) = \frac{\left[\sum_1^n (UR_i \times a_i)j\right] \times T_j}{\left[\sum_1^n (NR_i \times a_i)\right] \times (T_j + T_{conf\_j})} \times 100\%$$

$$= \frac{\sum_1^n (UR_i \times a_i)j}{\left[\sum_1^n (NR_i \times a_i)\right] \times \left(1 + \dfrac{T_{conf\_j}}{T_j}\right)} \times 100\% \qquad (2.8)$$

where
   T_j is the execution time
   T_{conf_j} is the configuration time for mode M_j

If the mode execution time is much greater than the mode configuration time, $T_j \gg T_{conf\_j}$, Equation 2.8 becomes equal to Equation 2.4. In other words, the effectiveness of a non-RTR FCR organization directly depends on the ratio T_{conf_j}/T_j. Hence, a non-RTR FCR is an effective solution when *mode execution time is much longer than FCR reconfiguration time for this mode or vice versa.*

For example, if the execution time of mode M_j is 1000 times greater than the configuration time for this mode, $T_j = 1000 T_{conf\_j}$, then

$$\text{RUF}(M_j) = \frac{\sum_1^n (UR_i \times a_i)j}{\left[\sum_1^n (NR_i \times a_i)\right] \times \left(1 + \dfrac{T_{conf\_j}}{1000 T_{conf\_j}}\right)} \times 100\%$$

$$= \frac{\sum_1^n (UR_i \times a_i)j}{\sum_1^n (NR_i \times a_i) \times 1.001} \times 100\% = \sim \text{RUF}_{exe}(M_j)$$

In the case where the execution time of mode M_j is equal to the configuration time for this mode, $T_j = T_{conf\_j}$, then

$$\text{RUF}(M_j) = \frac{\sum_1^n (UR_i \times a_i)j}{\left[\sum_1^n (NR_i \times a_i)\right] \times \left(1 + \dfrac{T_{conf\_j}}{T_{conf\_j}}\right)} \times 100\%$$

$$= \frac{\sum_1^n (UR_i \times a_i)j}{\sum_1^n (NR_i \times a_i) \times 2} \times 100\% = \sim 0.5 \,\text{RUF}_{exe}(M_j)$$

In other words, resource utilization in this case is twice lower, as no resources are involved in data-processing during the configuration period. This overhead caused by the mode configuration time has the same effect as the hardware overhead in a DPR FCR. It appears as if physical resources reserved for reconfiguration in an RTR FCR have *transformed* in the temporal domain into reconfiguration time.

2.5.4 Dependence of the Amount of Temporally Partitioned Resources and Performance on Task Segmentation Granularity

In previous sections, the mode of operation was considered as a set of functions deployed in the form of function-specific dedicated hardware components or function-specific processing circuits. This approach permits a system to reach the highest performance. If the performance can be compromised, it is possible to execute task/mode functions sequentially. In other words, segment S_i will directly correspond to one function $F_i \rightarrow S_i$. In this case, each mode of operation will be executed in sequential manner. For example, mode M_1 will require the following sequence of actions:

1. Configuration of the FCR for $S_1 \rightarrow F_1$, execution of data from the datasource according to function F_1, and then storage of obtained results in the temporal data memory
2. Configuration of the FCR for $S_5 \rightarrow F_5$, execution of data from the temporal memory according to function F_5, and then storage of obtained results in the temporal data memory
3. Configuration of the FCR for $S_1 \rightarrow F_6$ and execution of data from the temporal memory according to function F_6, followed by the ending of the mode cycle

Figure 2.14 shows the process of task mode execution using temporal partitioning of FCR resources into function-specific segments. This figure shows that the mode execution process requires much more time when compared with previous cases where all functions associated with mode M_1 were executed simultaneously. However, the amount of required resources is also lower. When following this approach, the total amount of resources that should be included in the FCR is equal to the maximum amount of resources required for any function in any mode:

$$NR_i = \max\{UR_{ij}\}, \quad \text{where resource type } i = 1, 2, \ldots, n$$

$$\text{and mode number } j = 1, 2, \ldots, m$$

In our quantitative example, the maximum number of CLB resources is associated with function F_4, which requires 378 CLBs. However, the maximum

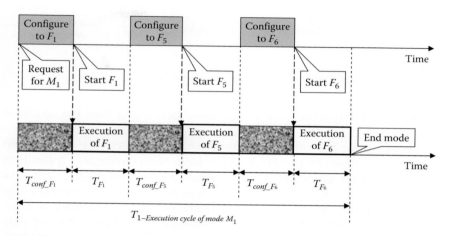

FIGURE 2.14
Mode M_1 execution process including functional component configuration times.

number of MACs and BRAM resources is associated with function F_1, which requires 42 MACs and 4 BRAMs (as per Tables 2.2 and 2.3).

Hence, the total amount of resources required for the FCR to execute any task mode is equal to

$$\sum_{1}^{n}(NR_i \times a_i) = 378 \text{ CLBs} \times 10 \text{ au} + 42 \text{ MACs} \times 100 \text{ au}$$

$$+ 4 \text{ BRAMs} \times 1000 \text{ au} = 11,980 \text{ au}$$

This represents only 47.6% of the resources required for an FCR where temporal partitioning of resources is oriented toward a complete set of mode functions (25,180 au).

In comparison with a DPR FCR, the reduction of resources is equal to 30,000 au/11,980 au = 2.5 times. The reduction in resources compared to an SFCR is equal to 37,200 au/11,980 au = 3.1 times.

In addition to the reduction of the total amount of resources in FCR, it is important to mention that any *task that uses the determined set of functions can run on such an FCR platform*. This platform allows temporal allocation of its resources to *one of the functions* from the aforementioned set. Hence, an FCR with limited hardware resources can accommodate a task where each mode requires higher amount of hardware resources than the existing FCR can provide. In this case, the task can be executed on an RCS function by function (from known set of functions) and thus become more general purpose than other classes of RCS. In such a system, each function is associated with a respective configuration bit-file. The task can be executed by sequentially loading and executing each task function, from first to last. The payment

Code of operation (Op-Code)	Source of operand 1	Source of operand 2	Place of result storage

FIGURE 2.15
Representation of an elementary task segment—*Instruction word.*

for adopting this approach consists of added processing time but no added hardware cost.

Moreover, since any function is associated with a certain algorithm and data structure, it can be presented in the form of an elementary (primitive) arithmetic or logic operators. Thus, the smallest segmentation for any task can be done at the level of elementary operations. These elementary segments should include one elementary operation (fine granularity of functions) associated with an elementary data structure. The elementary data structure for an elementary operation consists of one or two input data-elements (operands) and one resulting data element. This elementary segment can be encoded in the form of an *instruction* [17]. The general form of an instruction is shown in Figure 2.15.

The set of elementary functional segments can be represented by an *instruction set.* Execution of such an elementary segment takes relatively short time due to limited number of operations in hardware circuits. To be cost-effective, the reconfiguration time of an FCR to the next instruction also must be short (in the range of nanoseconds). This time limit dictates spatial allocation of all elementary functional components in FCR. The switching from one elementary function to another should be done by multiplexing the operational circuits: adders, logic circuits, shift registers, etc. Since most elementary operations are associated with arithmetic and logic functions, an FCR of this type can be considered as the ALU. The ALU is usually the operational core of the CPU, that is, the data-processor of conventional *instruction-based computers.*

The data-execution process in a CPU looks similar to the function execution process presented in Figure 2.14. The difference is that instead of executing functions F_i, the system will execute operations (elementary functions) O_i from the set of elementary operations $\{O_i\}$ $i = 1, 2,...,n$.

The reconfiguration time T_{conf_Oi} of the ALU from one elementary operation O_i to another, O_j, also will exist due to the following operating requirements:

1. Request of the instruction word and fetching said instruction
2. Decoding the instruction to get information regarding the operation code, associated sources of data, and destination of the result
3. Requesting and receiving the data from the addressed data source
4. Switching the ALU to the requested operation
5. Switching the ALU output to the addressed result storage

Thus, the architecture described earlier, which represents a conventional instruction-based computer (e.g., von Neumann or Harvard architecture), is the *ultimate form of RCS with temporal partitioning of resources* into elementary functions—arithmetic/logic operations. The reconfiguration of ALU architecture to accommodate the elementary function (instruction) is done by changing the topology of links between ALU components [18]. In this light, each *instruction can be considered as elementary configuration information* necessary for adaptation of the ALU architecture to one of elementary arithmetic, logic, or information exchange functions. Since the aforementioned reconfiguration is required for each elementary function, the ALU-based *elementary* RCS will be the slowest when compared with other RCS oriented toward more complex functions.

On the other hand, FCR using this type of *elementary* function execution will use minimum hardware resources compared with all other classes of FCR described earlier. That is why the actual cost of unused resources during reconfiguration periods is not high. Furthermore, the fine-grained nature of the CPU and sequential segment processing makes this class of systems the most flexible and, thus, the most general purpose of all RCS classes.

In other words, segmentation granularity directly influences on system performance, flexibility, and hardware/timing overheads associated with adaptation of computing architecture to different functionalities. Fine-grained segmentation of task functions allows respective reduction of hardware resources but increases the task execution time. On the other hand, coarse-grained segmentation increases the performance but requires more hardware resources and may cause higher hardware and timing overhead for system adaptation to the new function or task segment.

2.6 Summary

In this chapter, different aspects of the FCR organization were analyzed in general form using the RUF. This RUF was proposed for the evaluation of the effectiveness of various organizations of FCRs and RCS architectures. Now, it is possible to present the taxonomy of various FCR organizations in light of the discussed factors. This taxonomy is shown in the form of a classification tree in Figure 2.16.

The taxonomy shows that the most left-hand side variant of the FCR organization, fine grained (oriented toward elementary arithmetic/logic operations) with static configuration of all functional resources in the FCR area and temporal partitioning of resources into elementary task segments (instructions), can be considered the organization of a conventional ALU

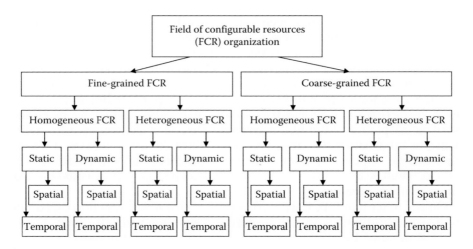

FIGURE 2.16
Taxonomy of organization of the field of configurable resources in reconfigurable computing
system.

in an instruction-based computer. This organization is the least dependant
on application specifics and, thus, is the most general-purpose computing
system organization. The drawback of this organization is the slowest data-
execution rate when compared to other FCR organizations using the same
hardware technology and clock frequency.

On the conceptually opposite end of the taxonomy, one finds the following
FCR organization: coarse grained, heterogeneous, and employing dynamic
reconfiguration of resources that were spatially allocated into large task segments (segments that are deployed only for specific modes of operation).
This type of FCR organization can provide the highest performance and
hardware resource utilization. However, this performance and RUF strongly
depend on the application details and on proper task segmentation for each
mode of operation.

All remaining FCR organization variants are found between these two
extremes and offer various compromises on performance and resource
utilization. Essentially, the more coarse grained the organization of the
FCR is, the higher the performance that can be reached using the same
amount of hardware resources. The same can be said when considering
heterogeneous versus homogeneous FCR organizations. As was shown in
Section 2.3, a heterogeneous FCR can provide better resource utilization
in the case where an application requires all types of resources included
in the FCR. Otherwise, certain types of resources will remain unused. On
the other hand, the dynamic reconfiguration of resources in the FCR can
also increase resource utilization if the task is multimodal and prediction
of upcoming modes can be done ahead of time. According to Section 2.4, if
the mode prediction time is greater than the reconfiguration time for this

mode, the reconfiguration of the FCR can be performed *seamlessly* from the point of view of the data-execution process. In general, dynamic reconfiguration increases resource utilization but depends on certain specifics of the application.

Temporal partitioning of resources into task functions discussed in Section 2.5 can reduce the system dependence on specifics of the application and can allow the computation of tasks that require more resources than the FCR can provide at a given time. In this way, hardware resources are transformed into a time cost. As a result, this transformation results in reduced system performance. On the other hand, utilization of resources can be better than in the case of spatial partitioning with dynamic reconfiguration due to the exclusion of the *dynamic hardware overhead*.

Generally speaking, the aforementioned taxonomy shows an increase in system performance and RUF as the FCR organization progresses from the left side to the right in the graph shown in Figure 2.16. However, this increase can only be obtained if the FCR organization reflects the characteristics of the application being implemented.

Exercises and Problems

Exercises

1. List two most fundamental aspects to be considered for synthesis of the reconfigurable system architecture.
2. What are the pros and cons of using fine-grained basic logic elements in the field of configurable resources?
3. Describe the organization of the configurable logic block and the fundamental building block of the field of configurable resources.
4. What is the main architectural difference between CPLDs and FPGA devices?
5. What is the difference between fine-grained and coarse-grained RCS architecture?
6. What are the main reasons for heterogeneous organization of the FCR in RCS? Define the pros and cons of heterogeneous FCR when compared to homogeneous FCR organization.
7. List the levels of FCR organization in RCS. Define the purpose of each level.
8. What is the main difference between statically configurable and statically reconfigurable FCR? Determine the difference in RUF calculation for the aforementioned classes of FCRs.
9. What is the main difference between statically and dynamically reconfigurable FCR? Determine the difference in RUF calculation for the aforementioned classes of FCRs.

10. What are the reasons causing *dynamic resource overhead* in the case of statically configurable and dynamically reconfigurable FCR organizations?
11. Determine the main difference between spatially and temporally configurable FCR.
12. What are the two main conditions that should be satisfied for *seamless* (noninterrupted) mode switching in dynamically reconfigurable FCR?
13. What is the main reason why statically reconfigurable FCR needs fewer resources than dynamically reconfigurable FCR?

Problems

Problem 1: To analyze the effectiveness of statically configured RCS, calculate the RUF for the FCR that consist of three types of resources:

R_1: CLB, each of which requires 16 area units (au)

R_2: MAC units as DSP components that need 128 au

R_3: Blocks of embedded random access memory (BRAM) that require 1024 au

The entire FCR contains 1024 CLBs, 128 MACs, and 16 BRAMs.

There are two tasks to be simultaneously executed on the RCS with the following requirements for resources:

Task 1 needs the following resources: 512 CLBs, 32 MACs, and 4 BRAMs.

Task 2 needs the following resources: 334 CLB slices, 62 MACs, and 8 BRAMs.

Calculate the RUF according to Equation 2.3.

Problem 2: For the aforementioned Task 1, it is possible to replace each MAC by 16 CLB slices. How will the RUF change in this case?

Problem 3: It is necessary to add the Task 3 to existing workload. The Task 3 needs the following resources: 76 CLB slices, 38 MACs, and 4 BRAMs. Taking into account that each MAC can be substituted by 16 CLBs, calculate the RUF value for SFCR.

Problem 4: According to the illustrative example discussed in Section 2.4.3, add the new mode of operation M_5. The specifications of new modes of operation are as follows:

1. M_5 can be initiated after the mode M_1 and followed by M_3.
2. M_5 consists of functions F_1, F_7, and F_8.
3. Specification of resources for these functions is provided in the following table.
4. Mode M_5 active time is up to 10 video-frames, that is, 0.333 s.

Functions of the Application	Resources Required for Function Implementation			Configuration Time for Function (ms)
	CLBs	**MACs**	**BRAMs**	
F_7	164	14	1	4.04
F_8	128	18	1	4.08

Calculate the following:

1. The RUF for the dynamically reconfigurable FCR performing all modes considered in the example with additional mode M_5 and according to Equation 2.6
2. Total amount of resources in the statically configurable computing system performing all five modes of operation and according to the denominator of Equation 2.3
3. The RUFs for the SFCR and dynamically reconfigurable FCR for the mode M_5 and according to Table 2.1
4. Reconfiguration time for mode M_5 (equal to T_{conf_5}) and reconfiguration time for M_3 (equal to T_{conf_3}) in case of partial reconfiguration of the FCR and according to Tables 2.1 and 2.3

References

1. A. B. Kahng, J. Lienig, I. L. Markov, and J. Hu, *VLSI Physical Design: From Graph Partitioning to Timing Closure*, Springer, Dordrecht, the Netherlands, 2011.
2. M. Xu, G. Grewal, and S. Areibi, StarPlace: A new analytic method for FPGA placement, *Integration—The VLSI Journal*, 44(3), 192–204, June 2011.
3. K. Vorwerk, A. Kennings, and J. W. Greene, Improving simulated annealing-based FPGA placement with directed moves, *IEEE Transactions on Computer Aided Design of Integrated Circuits and Systems*, 28(2), 179–192, 2009.
4. R. Hartenstain, A decade of reconfigurable computing: A visionary retrospective, in *Proceedings of the Conference on Design, Automation and Test in Europe (DATE 2001)*, IEEE Press, Piscataway, NJ, 2001, pp. 642–649.
5. S. Brown and Z. Vranesic, *Fundamentals of Digital Logic with VHDL Design*, 2nd ed., McGraw-Hill, New York, 2005, Chapter 3, pp. 73–136.
6. E. I. Muehldorf, Time split array logic element and method of operation, US Patent US3987286 A, Filing date: December 1974, Publication: October 19, 1976, Original Assignee: IBM Corp.
7. D. W. Page and L. R. Peterson, Re-programmable PLA, US Patent US4508977, Filing January 11, 1983, Publication: April 2, 1985, Original Assignee: Burroughs Corp.

8. N. Cavlan, Field programmable logic array, US Patent US4422072A, Filling July 30, 1981, Publication December 20, 1983, Organization Assignee: Signetics Corp.
9. Xilinx, *The Programmable Logic Data Book*, Xilinx Inc., San Jose, CA, 1999, Chapter 5.
10. R. H. Freeman, Configurable electrical circuit having configurable logic elements and configurable interconnects, US Patent US4870302 A, Filing date: February 1988, Publication: September 26, 1989, Original Assignee Xilinx Inc.
11. S. Vassiliadis and D. Soudris (Eds.), *Fine and Coarse-Grain Reconfigurable Computing*, Springer Inc., Dordrecht, The Netherlands, 2007.
12. M. B. Gokhale and P. S. Graham, *Reconfigurable Computing: Acceleration with Field-Programmable Gate Arrays*, Springer, Dordrecht, the Netherlands, 2005, Chapter 2.
13. A. M. Smith, Exploration of heterogeneous reconfigurable architectures, in *Proceedings of the 15th International Conference on Field Programmable Logic and Applications (IEEE FPL-2005)*, Tampere, Finland, August 2005, pp. 719–720.
14. S. W. Smith, *Digital Signal Processing: A Practical Guide for Engineers and Scientists*, Elsevier Science, Amsterdam, the Netherlands, 2003, Chapter 7.
15. M. Platzner, J. Teich, and N. When (Eds.), *Dynamically Reconfigurable Systems. Architectures, Design Methods and Applications*, Springer, Dordrecht, the Netherlands, 2010, Chapter 3.
16. C. Bobda, *Introduction to Reconfigurable Computing: Architectures, Algorithms and Applications*, Springer, Dordrecht, the Netherlands, 2007, Chapter 7.
17. D. A. Patterson and J. L. Hennessy, *Computer Organization and Design: The Hardware/Software Interface*, 3rd ed., Elsevier/Morgan Kaufman Publishers, Dordrecht, the Netherlands, Chapter 2.
18. D. A. Patterson and J. L. Hennessy, *Computer Organization and Design: The Hardware/Software Interface*, 3rd ed., Elsevier/Morgan Kaufman Publishers, Dordrecht, the Netherlands, Chapter 5.

3

Architecture of the On-Chip Processing Elements

3.1 Introduction

In Chapter 2, the ways of organization of the entire field of configurable resources (FCR) were considered. This chapter is focused on architecture organization of configurable or programmable processing elements (PEs). First, the organization of on-chip PEs will be discussed.

As per the definition of architecture given in Chapter 1, computing architecture can be described as the set of functional components, links between these components (spatial domain of the architecture), and procedures associated with components and links (temporal domain of the architecture). Thus, configuration of the architecture in the FCR assumes two levels:

1. Functional configuration and/or procedural programming of uniformed PEs, which customizes these PEs as functional components associated with current task mode for the period of activity of this mode of operation

2. Structural configuration of the FCR, which creates communication links between PEs, memory units, and input/output (I/O) interfaces according to dataflow dependencies in the task algorithm and its data structure

Thus, this chapter is oriented on the description of the first level of FCR, the on-chip PEs: fine-grained, coarse-grained, and hybrid PEs. Three major mechanisms for this functional configuration will be described and discussed on the conceptual level of consideration:

1. Configuration of PE by structural programming of interconnects between logic gates composing the PE

2. PE configuration based on functional programming of look-up tables (LUTs) and/or special mode registers

3. Procedural programming of CPU-based PEs

The pros and cons of each of the aforementioned mechanisms will be considered according to the type of PE granularity and types of operations.

3.2 Architecture of the Fine-Grained Configurable Processing Elements

From a general point of view, there are two ways to execute functional operations:

1. Get the result of the operation by data-execution according to the given function.
2. Retrieve the preprocessed result from the table of results according to the address. The address in this case can be equal to the value of source data.

The first method requires construction of the data-processing circuit from building blocks. This way assumes combining the logic circuits from logic gates according to given logic functions [1].

The second approach is memory centric and can be built around LUT. Both approaches are used in the on-chip level of FCR based on fine-grained logic blocks in complex programmable logic devices (CPLDs) and field programmable gate array (FPGA) devices [2]. To understand the aforementioned two concepts of implementation of the function-specific digital circuits, let us consider an example of some logic functions.

If the Boolean expression of the logic function is $\overline{A*B} + \overline{A}*C = Y$, then the truth table of this function will be as it is shown in Figure 3.1a. This circuit

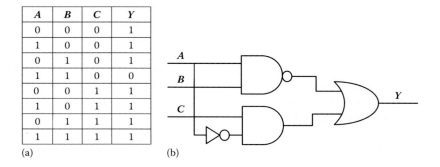

A	B	C	Y
0	0	0	1
1	0	0	1
0	1	0	1
1	1	0	0
0	0	1	1
1	0	1	1
0	1	1	1
1	1	1	1

(a)   (b)

FIGURE 3.1
(a) Truth table for logic function $\overline{A*B} + \overline{A}*C = Y$. (b) Logic circuit for the function $\overline{A*B} + \overline{A}*C = Y$.

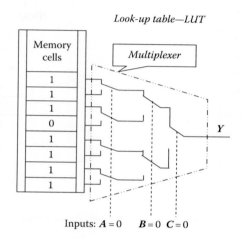

Look-up table—LUT

Inputs: $A = 0$ $B = 0$ $C = 0$

FIGURE 3.2
Implementation of the function $\overline{A} * B + \overline{A} * C = Y$ in a form of look-up table.

can be implemented using actual logic gates: AND, NAND NOT, and OR by certain interconnects as depicted in Figure 3.1b.

Alternatively, the same function can be implemented around LUT based on the array of 8 memory cells and multiplexer with 8 inputs to 1 output (8-to-1 MUX) as shown in Figure 3.2.

Indeed, in this way of implementation, all 8 possible output results for the 3-input logic function are stored in the eight 1-bit-wide memory cells (8 flip-flops).

The inputs for this logic unit are selectors: A, B, and C of the 8-to-1 multiplexer. And the output of this multiplexer is the output of the LUT. Figure 3.2 depicts the case when 0 values are applied to all inputs A, B, and C.

The content of memory cells reflects the truth table content for this logic function. Therefore, only in case when $A = 1$, $B = 1$, and $C = 0$, the position of switches in the multiplexer allows connecting the output Y to the memory cell storing 0 bit. In all other cases, Y output will be connected to memory cell storing value of logic "1."

Let us now consider the case when the functionality of the logic circuit should be changed for the new mode of operation.

For example, the new logic function $\overline{A} + \overline{B * C} = Y$ has to be configured on the basis of the same logic gates: AND, NAND NOT, and OR. The truth table of this function is shown in Figure 3.3a. And the logic circuit for this function is presented in Figure 3.3b.

To reconfigure these logic gates from one functional configuration shown in Figure 3.1b to the new configuration shown in Figure 3.3b, a certain switching infrastructure is needed in case an actual logic circuit should be

A	B	C	Y
0	0	0	1
1	0	0	1
0	1	0	1
1	1	0	1
0	0	1	1
1	0	1	1
0	1	1	1
1	1	1	0

(a)   (b)

FIGURE 3.3

(a) Truth table for logic function $\overline{A} + \overline{B * C} = Y$. (b) Logic circuit for the function $\overline{A} + \overline{B * C} = Y$.

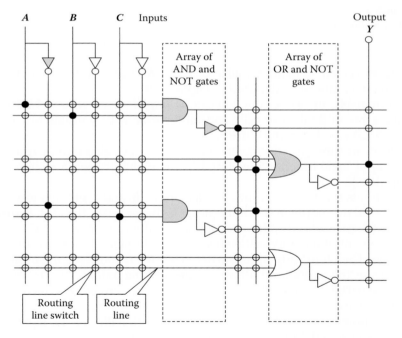

FIGURE 3.4

The example of the framework for the configurable logic circuits.

used for logic operations [2]. As an example, let us consider the framework for the configurable logic circuits presented in Figure 3.4.

This framework includes three digital inputs: *A*, *B*, and *C*, arrays of AND, NOT, and OR logic gates, and an interconnection network that consists of routing lines and routing line switches located in every cross of vertical and

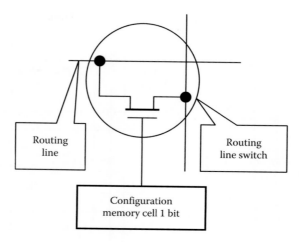

FIGURE 3.5
Cross-point programmable switch controlled by 1-bit configuration memory cell.

horizontal routing lines. Thus, this network makes possible to connect any input to any logic gate directly or inverted. In addition, this network allows connection of any logic gate output to the output line (*Y*) directly or inverted. This framework is the hardware base of programmable logic array (PLA) devices or programmable logic devices (PLDs) [3].

The routing line switches usually are based on pass transistors as shown in Figure 3.5. If the control line of this switch is connected to the 1-bit-wide configuration memory cell, the interconnection between vertical and horizontal routing lines can be controlled by the content of the associated configuration memory cell. For example, if this memory cell contains logic "1," the switch is closed and thus the signal from one routing line can go through the switch to the other line. If configuration memory cell stores the bit $= 0$, then the switch is open and the associated horizontal and vertical routing lines are isolated. Therefore, interconnection between two routing lines is programmable, and the routing line switch usually is called *cross-point programmable switch*.

Let us depict all closed switches in black and transparent if they are open. In this case, Figure 3.4 presents the configuration of interconnection network implementing logic circuit for the function $\overline{A} * B + \overline{A} * C = Y$ according to the schematic of this circuit presented in Figure 3.1b. In Figure 3.4, input *A* is directly connected to the first (upper) AND logic gate and over the inverter (NOT) to the second (lower) AND gate. Points of interconnections are shown as black "dots" (closed cross-point switches). All other interconnects are shown in the same manner. Since each routing switch is controlled by 1-bit memory cell, there should be 72 bits of information associated with 72 cross-points (8 rows $\times$ 9 columns). This number of routing switches are enough to configure logic function, which requires up to 2 AND or NAND gates plus 2 OR or NOR gates.

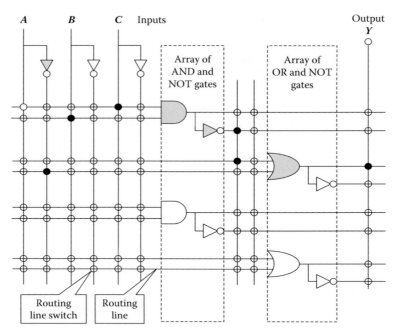

FIGURE 3.6
The example of the configuration of the logic circuits performing function $\overline{A} + \overline{B} * C = Y$.

For example, if it is necessary to switch from the function $\overline{A} * B + \overline{A} * C = Y$ to the function $\overline{A} + \overline{B} * C = Y$, the content of configuration memory should be changed to activate a different set of switches, as shown in Figure 3.6.

Thus, downloading the new content of configuration data to the configuration memory allows changing the circuit architecture and switch from one functionality of the logic circuit to another.

In other words, this approach makes possible to change the system functionality by programming the circuit configuration (in spatial domain) but not the procedure (temporal domain).

However, it is necessary to mention that the percentage of used gates and routing switches never reaches 100% as it can be in application-specific logic circuit (ASIC) designs. Indeed, in the first function shown in Figure 3.4, the number of used gates (AND, OR, and NOT) is 5 out of 11.

The resource utilization factor $\text{RUF}_{gates}(F_1) = 5/11 \times 100\% = 45.5\%$.

And for the second function shown in Figure 3.6, $\text{RUF}_{gates}(F_2) = 4/11 \times 100\% = 36.4\%$.

Utilization of routing resources consisting of 72 cross-point programmable switches and associated 72-bit configuration memory is as follows:

For the function F_1, $\text{RUF}_{sw}(F_1) = 9/72 \times 100\% = 12.5\%$.

For the function F_2, $\text{RUF}_{sw}(F_2) = 6/72 \times 100\% = 8.33\%$.

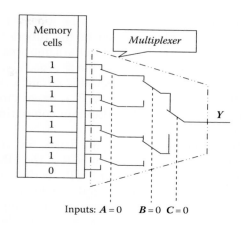

Inputs: $A = 0$ $B = 0$ $C = 0$

FIGURE 3.7
Implementation of the function $\overline{A} + \overline{B} * C = Y$ in look-up table.

Nevertheless, this approach was widely used in different PLDs and then in CPLDs in the form of macrocells.

Let us now consider the implementation of the function-specific digital circuit built around the LUT. In case the functionality of the circuit needs to be changed from the first logic function $A * B + \overline{A} * C = Y$ to the next function $\overline{A} + \overline{B} * C = Y$, the only necessary thing to do is to reprogram the LUT memory with new content according to the new truth table associated with new logic function. In our example, the content of LUT memory is shown in Figure 3.7.

Comparing this figure with Figure 3.2, it is possible to see that there is no change in hardware schematic of the circuit—only the content of memory cells is different.

It is interesting to compare the required resources for the function-specific circuit based on the LUT with the same circuit based on the arrays of actual logic gates. As mentioned earlier, the function-specific circuit based on PLAs required 72-bit configuration memory (e.g., Figure 3.4). The LUT-based circuit with the same functionality requires only 8-bit memory for programming any logic function with 3 inputs and 1 output.

Furthermore, the LUT multiplexer needs seven single-pole, double-throw (SPDT) switches. In the case of MOSFET implementation, it requires twice more transistors than in the case of PLAs. Thus, instead of 72 switches for routing lines in the case of PLA, the LUT switching scheme (multiplexer) requires five times less equivalent switches—$7 \times 2 = 14$.

At the same time the RUF, in the case of LUT implementation, is 100% because for any logic function implemented in LUT-based logic circuit, all memory cells as well as all switches in the multiplexer will be used. Moreover, there are no actual logic gates to perform the Boolean function execution, and thus, no logic resources must be spent for that. In turn, the LUT-based approach for implementation of the function-specific circuits provides higher density of logic deployed in the same area of the FCR.

However, there is a certain drawback in the approach based on LUTs. The response time for LUT-based function-specific circuit is somewhat higher in comparison to the response time of the PLA-based circuits. The reason is longer switching time in the LUT multiplexer comparing to propagation delay in actual logic gates.

Nonetheless, this approach became the basis for configurable logic blocks (CLBs), the main building blocks in FCR of the FPGA devices. Most of FPGAs are using from 4- to 6-input LUTs in their CLBs. Obviously, LUTs with 4 inputs require 16 memory cells for logic function programming when 6-input LUTs need 64 memory cells of configuration memory. Certainly, 6-input LUTs can accommodate more complex logic functions than 4-input LUTs. As the trade-off the 6-input LUT can be divided on two 5-input LUTs.

From general point of view, increasing the number of LUT inputs increases complexity of logic functions but decreases the performance parameters. Indeed, for N-input LUT, multiplexer will require $(2^N - 1)$ SPDT switches. The more switches involved in accessing one of the memory cells, the longer is LUT memory cell access time. On the other hand, reducing the dimensions of the MOSFET transistors reduces internal capacitance and thus reduces switching time. Hence, negative aspects of increasing the number of LUT inputs could be mitigated (until certain level) by reduction of transistor dimensions. For example, Xilinx Virtex-7 family of FPGA devices that uses 28 nm MOSFET technology provides CLBs with 6-input LUTs [4]. However, each 6-input LUT can be reconfigured as two 5-input LUTs. In other words, the LUT with 6 inputs require 64 memory cells, but being used as two 5-input LUTs, uses 32×1 bit memory cells for each LUT. At the same time, the 6-input MUX containing 63 SPDT switches can be divided on two 5-input MUXs with 31 switches. This allows flexible utilization of configuration memory and MUX resources according to application needs.

Let us now consider the performance of LUT MUX in terms of timing. For simplicity, we will take the 2-input LUT with 4 memory cells and multiplexer consisting of 3 SPDT switches depicted in Figure 3.8. MUX switches S_1, S_2, and S_3 may have different switching time caused by the manufacturing process and other factors. This timing also depends on the operation temperature.

For example, when binary digit 11 is applied to inputs $A = 1$ and $B = 1$, switches S_1, S_2, and S_3 are starting to change their positions from the initial position depicted in Figure 3.8. Thus, the output $Y = 0$ in this moment of time. This state is shown in Figure 3.9 at T_0.

When switching process has started, all switches in the LUT MUX are in transition process. Being not absolutely identical (e.g., a small difference in

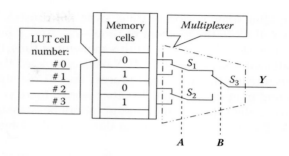

FIGURE 3.8
A 2-input look-up table with 3 single-pole, double-throw switches S_1, S_2, and S_3.

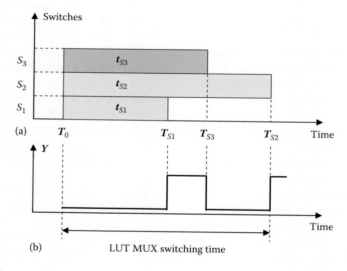

FIGURE 3.9
Switching process in look-up table multiplexer. (a) Transition periods for switches S_1, S_2, and S_3 and (b) variation of MUX output value during switching time.

the area of transistors, variation in gate capacitance), switches S_1, S_2, and S_3 may have different transition periods: t_{S1}, t_{S2}, and t_{S3}.

Let us assume that the switch S_1 reacts faster than other switches and connects its "pole" to the output of the memory cell #1 at the time T_{S1} as shown in Figure 3.9a. In this moment of time, the MUX output $Y = 1$ because switch S_3 still is connected to the "pole" of S_1, and thus, the output Y is connected to the memory cell #1 of the LUT via switches S_1 and S_3.

Then, in the moment of time T_{S3}, switch S_3 connects its "pole" to the output of switch S_2. Being still connected to the memory cell #2, switch S_2 transmits the value of logic 0 stored in this cell. Thus, the MUX output now shows

$Y=0$. However, after some period of time, switch S_2 finishes its transition to the memory cell #3 in the moment of time—T_{S2}. And Y becomes equal to logic 1 the value stored in cell #3. At this moment of time, the process is over because all switches in the LUT MUX have completed their transition processes. Thus, during this transition process the MUX outputs different values in different moments of time. In other words, there is certain oscillation period when the value of MUX output cannot be considered as valid. In general case, the more switches are in the MUX, the longer oscillation period could be.

Therefore, it is necessary to wait some determined period before reading the valid LUT output. However, during this period, the value of inputs A and B may also change. This may consequently change the LUT output again. Hence, to provide deterministic performance to the LUT-based logic circuit, the inputs must keep their values for the longest period of oscillations in LUT and the output value should be registered in certain moment of time. Thus, the period of time longer than LUT MUX switching time should be determined as *the clock period*. Furthermore, this *clock period should be greater than maximum period of signal propagation from their sources to the inputs of the LUT plus LUT MUX switching time*. This condition should be satisfied in full range of operational temperatures possible for the circuit. Otherwise, the aforementioned uncertainty in output of the LUT-based logic element can occur and the system will not operate properly. Therefore, the aforementioned LUT MUX output should be connected to the flip-flop data input with clock signal used as the strobe. This circuit is depicted in Figure 3.10.

This additional D flip-flop (DFF) and proper selection of clock period provide deterministic performance of this circuit and eliminate problems associated with the time difference of the input signal arrival and the non-identical switching time of transistors in the LUT MUX. The example of the aforementioned process for the LUT-based circuit shown in Figure 3.8 is presented in Figure 3.11.

This figure shows that signals coming to the inputs A and B have different propagation periods—t_A and t_B. This may happen because of several reasons:

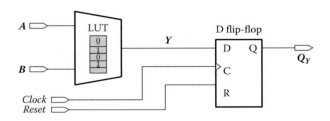

FIGURE 3.10
Look-up table with D flip-flop as 1-bit output register.

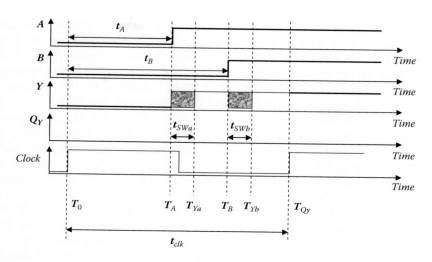

FIGURE 3.11
Mitigation of different signal propagation time and transition process in look-up table.

1. Different distances (length of routing lines) from the sources of the aforementioned signals
2. Difference in capacitance and/or inductance of the routing lines conducting signals
3. Difference in electrical characteristics of output circuits generating these signals

After the arrival of each input signal, there is a certain period of LUT MUX switching time, t_{SW}, considered in Figure 3.9. Thus, between arriving moment of a signal (e.g., T_A for input A) and moment of time when LUT output, Y, produces a valid level of output signal is the period of MUX switching time—t_{SW}. Hence, to avoid influence of the aforementioned unpredictable differences of signal propagation time and LUT MUX switching time, the minimum clock period, T_{clk}, should be greater than maximum signal propagation time plus maximum LUT switching time:

$$\min\{t_{clk}\} = \max\{t_A, \ t_B\} + \max\{t_{SWa}; \ t_{SWb}\} \tag{3.1}$$

Indeed, if rising edge of the clock is used as a strobe for generating signals coming to the inputs A and B in a moment, T_0, and registering the LUT output in the DFF in a moment, T_{Qy}, all uncertainties in status of the LUT output cannot affect the performance of the system. This is true because in the moment of next rising edge of the clock T_{Qy}, all transition processes in MUX have already completed and input signals propagation times are within the clock period as well.

The problem associated with the elimination of input signals propagation delays is also valid for the reconfigurable logic circuits based on PLAs. Despite the fact that there are no uncertainties caused by switching circuits (as at LUT MUX), there is a difference in signal propagation time inside the routing matrix and pass transistors in PLA. Furthermore, this difference depends on routing configuration and thus may vary. Hence, for different routing configurations (e.g., shown in Figures 3.4 and 3.6), signal propagation delay in the routing matrix may be different. Hence, the same effect as depicted in Figure 3.9 can occur. Therefore, the same approach with additional flip-flop as the PLA output register can be considered as a solution for this case too. An example of the organization of PLA-based reconfigurable logic circuit with DFF is shown in Figure 3.12.

On the other hand, the condition for the minimum clock period reflected in Equation 3.1 is one of the limiting factors for the performance of the RCS built on the basis of the aforementioned fine-grained reconfigurable logic circuits. Indeed, in the dedicated ASICs (e.g., ASICs), all routing lines between logic gates are predetermined and adjusted in timing on the design stage. Therefore, the difference in signal propagation delays can be minimized. Elimination of routing switches and/or multiplexers excludes the respective uncertainties in switching time and signal propagation times as well. In turn, the clock frequency in the ASIC can be increased, and thus, ASIC can provide higher performance for the same task when compared with LUT-based or PLA-based logic circuits. The downside of the ASIC-like computing circuits is the lack of flexibility.

Thus, the aforementioned reduction of clock frequency in the reconfigurable logic blocks is the timing cost paid for higher functional flexibility. Nevertheless, the balance between flexibility and in-circuit programmability versus reduced operational frequency made the aforementioned LUT-based CLBs the main building blocks in FPGA devices. Definitely, CLBs contain some additional circuits that allow integrating them into larger PEs. For example, CLB may contain 1-bit adder circuit with carry-in and carry-out bit (C-in and C-out). This makes a possible composition

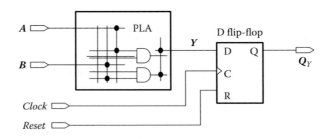

FIGURE 3.12
Reconfigurable programmable logic array–based circuit with D flip-flop as 1-bit output register.

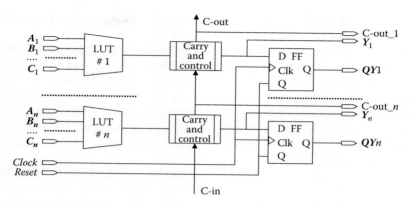

FIGURE 3.13
General organization of multi-look-up table–based configurable logic block as the processing element of the field of configurable resource.

of multibit adders by cascading several CLBs. Since arithmetic and logic operations in most cases are multibit, in modern FPGA devices the CLBs consist of several slices, each of which is combined by the set of LUTs with associated flip-flop registers and carry and control logic. For example, in 7 series of Xilinx FPGA devices, each CLB is combined by 2 slices, each of which contains four 6-input LUTs and 8 flip-flops and chain of carry-in/carry-out bits [4]. This is an effective technique when multibit PEs should be composed in the FCR. Being deployed near each other, several LUTs with associated flip-flops in the same slice can exchange the information using local and therefore short communication lines. In turn, this reduces signal propagation time and allows increasing the operational clock frequency.

The general organization of multi-LUT CLB as the multibit PE in the fine-grained FCR is shown in Figure 3.13.

In case of PLA-based PEs, their composition also assumes deployment of PLAs into larger macroblocks. These architectures of multibit PEs also became the de facto standard components in the CPLDs [5].

3.3 Architecture of the Coarse-Grained Configurable Processing Elements

As mentioned in Chapter 2, any processing architecture can be composed only by fine-grained PEs based on CLBs. However, the efficiency in resource utilization is higher when FCR is heterogeneous. The heterogeneous FCR should include some types of function-specific processing units (FPUs) as shown in Figure 2.2.

Since most of the applications require more complex operations than just bit-wide logic operations, word-wide multipliers, adders, and shift registers usually are needed. It is necessary to mention that shift registers can perform multiplications or divisions on 2^n by shifting the binary word to the left (multiplication) or right (division) n-shifts. Recently, these shift operations can be done using LUT memory as shift register (from 16- to 128-bit shift registers [4]). In this case LUT-based PE can be reconfigured to the shift register without any additional flip-flops.

For general word-wide operations, it would be useful to have in the FCR the FPUs dedicated for these operations. As an example for the aforementioned FPU, the multiply-accumulate (MAC) multimodal unit can be considered. All modern FPGAs (e.g., Xilinx 7 series FPGAs) are equipped with such FPUs. These FPUs can perform word-wide addition/subtraction as premultiplication operation, integer multiplication, and postmultiplication accumulation of results. Registers in between these operations allow pipelining of add/subtract/multiply operations that could accelerate the computational process [6]. This is specifically important for macro-operations associated with digital signal processing (DSP) functions (e.g., infinite impulse response [IIR] and finite impulse response [FIR filtering], fast Fourier transform [FFT] [7]). Therefore, these types of FPUs were called in Xilinx FPGAs as DSP slices. The general organization of MAC FPU is depicted in Figure 3.14. This type of FPU may be configured for different sequences of functions that can be pipelined for performance acceleration. Here, the configuration memory stores the operation mode identifier—op-mode configuration word.

Each of the bits in this word configures certain component in the MAC FPU. Some of them are the determining type of operation (e.g., add or subtract) and some of them select the status of register multiplexers (e.g., store the result in the register or bypass the register). Thus, op-mode configuration word allows organization of pipelined, partially pipelined, or nonpipelined data-execution process with different functionalities of dedicated PEs. Definitely, these types of pipelined PEs are effective for computationally intensive tasks or segments of tasks where relatively large data should be processed by the same algorithm. In this case the number of bits in data-words and number of words to be executed at a time can be considered as massive input data. Therefore, the number of PEs and their bit resolution could be configured accordingly. For example, the ALU can perform a set of arithmetic or logic functions on two 32-bit operands. But if operands are 16 bits and ALU allows reconfiguration for 16-bit operations, then ALU can perform same operation on two pairs of 16-bit operands simultaneously. Thus, the performance can be doubled. This type of operation on reconfigurable ALU is shown in Figure 3.15a and b.

Same manner ALU performance can be quadrupled in case of 1-byte operands. In this case each 16-bit register is divided on two 8-bit registers and 32-bit ALU will be divided on four 8-bit ALUs. This is possible to do by

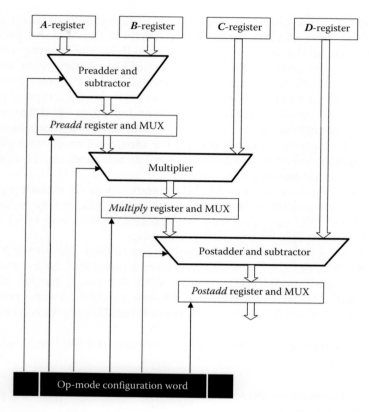

FIGURE 3.14
General organization of the multiply–accumulate function-specific processing unit.

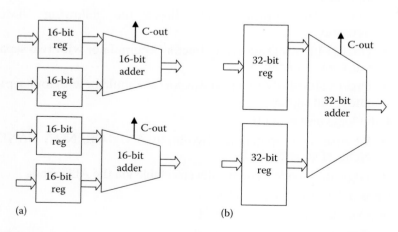

FIGURE 3.15
Reconfigurable ALU as parallel 16-bit adder (a) and as single 32-bit adder (b).

blocking carry-out (C-out) bit from the ALU previously processing a lower byte (of the 16 bit word) to the carry-in (C-in) bit of the ALU that processed a higher byte of the 16-bit word. This can be done just by programmable switches and associated bits programmed in the configuration memory. Hence, it is possible to reconfigure multimodal ALU to different processing schemes without much additional hardware overhead. This type of parallel processing often is called single instruction multiple data (SIMD) operations. However, here instead of *instruction* caring control information, the *configuration word* allows the ALU performing the same *function* on multiple data-words simultaneously. It would be clearer to call this mode of ALU operation as single function multiple data (SFMD).

Obviously, the aforementioned SFMD operation does not exclude pipelined process of data-execution. Hence, this multimodal ALU can be included in the pipelined FPU equipped with multiplier(s), shifters, and other coarse-grained function-specific processing circuits.

As an example it is possible to mention the DSP481 slice in the Xilinx 7 series of FPGA devices [10] where instead of a postadder device, the aforementioned multimodal ALU coupled with the result pattern detector is included. The basic functionality of this pipelined multimodal FPU equipped with reconfigurable ALU capable for SIMD operations is pictured in Figure 3.16.

In this example the DSP481 slice FPU includes the following functional components:

- 25-bit preadder/subtractor coupled with the result register for pipe-lined operations
- 25 × 18 bit multiplier coupled with a multiply register and a set of multiplexers
- 3-input adder/subtractor that allows the following SIMD configurations:
  - Dual 24-bit SIMD adder/subtractor/accumulator with two separate C-out signals
  - Quad 12-bit SIMD adder/subtractor/accumulator with four separate C-out signals
- 48-bit logic unit providing:
  - Bit-wise logic operations: two-input AND, OR, NOT, NAND, NOR, XOR, and XNOR
  - Logic unit mode dynamically selectable via the ALU mode
- Pattern detector that provides
  - Overflow/underflow support
  - Convergent rounding support
  - Terminal count detection and autoresetting

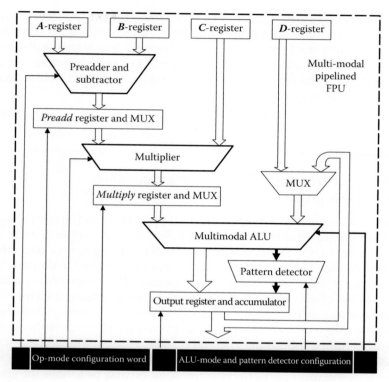

FIGURE 3.16
Pipelined multimodal function-specific processing unit with single instruction multiple data capable reconfigurable ALU.

- Configurable cascading circuits
- Dynamic user-controlled operating modes (using 7-bit op-mode control bus) (e.g., DSP481 slice in the Xilinx 7 series of FPGA [6])

The aforementioned unit was presented to illustrate real multimodal FPUs as the coarse-grained PEs oriented on integer operations on the streamed dataflow.

For many areas of application, the floating-point (FP) arithmetic operations are needed. It is possible to perform FP operations using the integer ALU/multiplier circuits. However, the effectiveness of resources utilization could be higher if special FP-FPUs are included into the FCR.

There could be many specific arithmetic/algebraic/trigonometric functions to be used in application. It may be necessary to know up front how often they are required for computation. If they are expected to be in use quite often, it may be practically useful to develop them as functional PEs and implement as coarse-grained integrated circuits in the FCR.

The concept assumed utilization of the coarse-grained application-specific hard cores as the base-uniformed PEs. This approach is used in many coarse-grained reconfigurable array (CGRA) systems such as eXtreme Processing Platform (XPP) [10] where architecture consists of several processing array clusters (PACs). The survey of architecture organization of CGRA systems can be found in many sources (e.g. [7]).

3.4 Architecture of the Hybrid Programmable Processing Elements

As discussed in Chapter 1, the best class of applications for the computing systems with programmable architecture (RCS) is class of computation-intensive tasks. In other words, RCS is effective for tasks where big data or data-streams should be processed in a relatively short time or in run-time. On the other hand, there are many tasks where many operations should be performed on relatively small number of data-elements without strict timing constraints. This class of applications usually is called algorithmically intensive. Generally speaking, any application may contain both algorithmic-intensive and computation-intensive parts in certain proportion. Therefore, for the cases where computation-intensive part is not a dominant segment of the task, but there are many algorithmic-intensive parts, the effective solution may be to have the general-purpose processor(s) as the PPEs coupled with the FCR. This class of RCS often was called *hybrid* RCS. Since there are several and quite different meanings of the term "hybrid," the definition of the hybrid RCS is needed.

Definition 3.1 *Hybrid RCS* is a system that consists of one or many general-purpose processor(s) coupled with the FCRs.

Thus, hybrid RCS consists of two types of PEs according to the classification given in Chapter 1, Figure 1.3: (1) PPEs based on the concept of computing systems with programmable procedure and (2) configurable processing elements (CPEs) based on the concept of computing systems with programmable architecture.

The PPE(s) can be allocated (1) on the same die as CPEs (embedded solution) or (2) outside the FCR. There are some advantages and disadvantages of both cases.

In this chapter the embedded system-on-chip (SoC) solutions will be analyzed. And the solution associated with external to FCR implementation of the PPE will be discussed in Chapter 5.

If PPEs are deployed on the same die (chip), the major advantages for the RCS are the following:

- *Acceleration of data transfer* between PPEs, CPEs (hardware accelerators), memory modules, and I/O interfaces. This is possible to do because of three reasons: (1) Distances between communication nodes are shorter than in the case of off-chip communication. Thus, the frequency in communication bus(es) or links can be higher than in the case of off-chip allocation of the PPEs. (2) A number of communication lines utilized for on-chip communication can be much higher than in the case of off-chip busses or links. (3) It is not necessary to use I/O buffers and multilayer onboard data-transfer lines. This allows reduction of system cost as well.

- *Customization of interface(s)* for the PPE. Indeed, the application may need to process the data from different sources and output them in real time to different actuators. In this case on-chip custom I/O drivers associated with dedication (to the set of sensors and actuators) may be a more effective solution in comparison to off-chip (onboard) application-specific data acquisition and outputting subsystems. Furthermore, the on-chip interface infrastructure could be adapted to new protocols or communication standards without redesigning the entire board. On the other hand, on-chip deployment of the custom interface(s) reduces the dimensions of the system, power consumptions/dissipation, and finally the cost of production/maintenance/transportation and storage.

- *Increasing system reliability.* This is possible because more components and links between components are deployed on the die; less components and associated lines, via(s), and soldered points are on the board. However, the aforementioned soldered points (pins of the chips, passive components, connectors, sockets, etc.) as well as via(s) in multilayer printed circuit boards (PCBs) are much more sensitive to vibration, climatic changes, manufacturing defects, etc., in comparison to on-chip components interconnects.

- *Reduction of manufacturing cost.* (1) Reduction of number of components to be ordered, shipped, stored, assembled, and tested; (2) reduction of the cost of PCB design, layout, production; (3) reduction of the board assembly time and cost; and (4) reduction of the cost of board testing, packaging, and shipping.

- *Increasing sustainability* of system production. This aspect is one of the most important in production. However, production sustainability depends on the availability of components. Any component discontinuation/replacement often requires partial or even entire redesign of the system. In case of on-chip deployment of the hard-core PPE, this aspect could be mitigated because the processing core and all associated interface circuits and drivers are allocated or configured inside the programmable chip (e.g., FPGA device).

On the other hand, there are certain drawbacks of the aforementioned organization of hybrid RCS. In case hard-core conventional processors (PPE) are
deployed in the same chip where the reconfigurable processing elements are
combined in the FCR, designers have certain limitations:

- *The number of PPEs* on-chip is always the same. If the application
  requires more processors for parallel execution of the algorithmic-
  intensive segments of the task, it is not possible to add more PPEs to
  the circuit. In this case these segments should be executed sequentially, which may reduce the performance of the system.

  On the other hand, if the number of PPEs on-chip is more than
  the application requires, there is a certain hardware and power overhead that reduces the cost-efficiency of the system.

- *The type of the PPE* and the associated communication/memory
  infrastructure may not satisfy application specifications and thus do
  not allow reaching the required performance. On the other hand,
  the complexity and cost of the embedded PPEs can be overkilling for
  the application and thus will reduce cost-efficiency.

- *The cost of the PPE* may be higher than the cost of the stand-alone
  off-the-shelf processor and the associated software development
  cost. This issue is just another aspect of the previous problem when
  application requirements may not fit architecture of the PPE embedded on the die with reconfigurable part of the system.

Nevertheless, the on-chip embedded PPEs (often called embedded processor
blocks or hard cores) became the de facto standard option for most FPGA
families provided by (1) Altera (e.g., advanced RISC machine (ARM)
Cortex-A9 in Cyclone-V [10]), (2) Xilinx (e.g., Power PC-440 in Virtex-5 [11]
or dual-core ARM Cortex-A9 in Xilinx Zynq-7000 FPGAs [12]), (3) Actel
(e.g. ARM® Cortex™-M3 in SmartFusion FPGAs [13]), etc. There are different architectures of the embedded SoC processors or hard processing
cores in different families of FPGA devices designed by different vendors.
Nonetheless, there are common components that represent the general architecture of the aforementioned on-chip general-purpose processing cores as
PPEs. The set of these common components consist but is not limited to

1. Single- or multicore general-purpose processor. The architecture of
   this processor is mostly reduced instruction set computing (RISC)
   Harvard-like architecture. Thus, instruction and data memory
   modules/cache are separated with separate instruction and data
   buses. Hence, most of processing cores are coupled with instruction
   cache and data-cache memories.

2. Communication infrastructure of the memory hierarchy. This consists of (a) internal (on-chip) memory bus, (b) direct memory access

(DMA) controllers, and (c) arbitration circuits and interface(s) to the external (main) memory.

3. I/O communication infrastructure. This consists of (a) internal (on-chip) peripheral bus, (b) controllers for standard I/O devices and ports, (c) and customizable ports for communication with application specific (custom) peripherals.

4. Debugging circuits and port(s).

As an example, it is possible to consider the hard-core processor in the Xilinx Zynq-7000 platform devices [12] mentioned earlier. The platform consists of two parts:

1. Heterogeneous FCRs, called "programmable logic." This part includes LUT–based CLBs as fine-grained elements, DSP-oriented coarse-grained blocks, and on-chip RAM blocks similar to discussed in Sections 3.2 and 3.3.

2. Hard-core general-purpose processor based on ARM Cortex-A9 superscalar architecture. This dual-core architecture consists of two processor cores, each of which is coupled with individual 32 kB instruction and 32 kB data-caches (32/32 kB I/D caches) L1 level. These cache modules are exchanging data with the common 512 kB cache L2 level. On the other side both processing cores are tightly connected to individual DSP and FP units that work as the function-specific units (hardware accelerators). This dual-core processor is equipped with general interrupt controller, watchdog timer and set of other timers, DMA controller and snoop control unit for cache coherency. The ARM architecture is well presented in the literature (e.g., [14]).

In addition to this Xilinx Zynq-7000 provides the set of embedded interfaces to external devices such as (1) serial peripheral interface (SPI); (2) inter-integrated circuits I2C interface; (3) controller area network (CAN), universal asynchronous receiver/ transmitter (UART); and (4) universal serial bus (USB). These interfaces are de facto standard for most embedded microcontrollers and embedded hard-core processors providing interfaces to the low-bandwidth external I/O devices. The built-in gigabit Ethernet allows interfacing with external high-bandwidth devices and networks. All the aforementioned interfaces could be activated and configured via dedicated registers and then multiplexed to the actual I/O pins via special multiplexer—processor I/O MUX.

As it could be seen, the on-chip hard-core processor can provide a wide variety of on-chip resources associated with interfacing to the memory devices, different peripherals, and networks. This allows simplification of the design process and reliable inclusion of the SoC to the RCS in all levels: (1) on-chip via common bus (e.g., advanced microcontroller bus architecture (AMBA) with advanced eXtendable interface (AXI) ports [12]), (2) onboard level via general-purpose input/output as custom parallel interface and SPI and I2C as standard serial interface, (3) on-system level via CAN, UART, USB standard interfaces, and multigigabit transceivers as well, and (4) intersystem level of communication using network interfaces such as USB and Ethernet.

In addition to the aforementioned, relatively rich memory hierarchy often with ability for virtual memory organization makes the on-chip hard-core processor a very valuable unit in such hybrid RCS architecture. The engineering practice shows that even increased cost of the FPGA with embedded PPE may reduce system cost because of reduced design and manufacturing time remote debugging, servicing, and often remote restoring the RCS functionality. Therefore, the embedded PPEs becoming more and more popular design solutions in the RCS architectures.

3.5 Summary

This chapter described the architecture organization and methods for functional customization of the on-chip PEs of the FCR.

There are different methods for implementing the arithmetic–logic functions in the fine-grained PEs and complex macrofunctions in the coarse-grained PEs. The most complicated application-specific macrofunctions usually require hybrid organization of PEs. These types of PEs contain the general-purpose RISC processing core(s) combined with application-specific hardware accelerator(s) composed of fine-grained and coarse-grained PEs. The general-purpose processing cores are instruction processors implemented in the form of CPU-centric unit(s) embedded to the FPGA platform.

In case of fine-grained PEs, there are two methods for programming ALU functions into the PE:

1. Method based on structural configuration of links between logic gates inside each PE.
2. Method based on programming the LUT in the PE. The inputs of this type of PE work as the control signals for the LUT multiplexer that provides connection of the output to the appropriate memory cell of the LUT.

The first method is used for PEs (called macrocells) in the FCR deployed in CPLDs. The second method became a de facto standard in the FCR of the FPGA devices.

In contrast to the fine-grained PEs, the coarse-grained PEs are configured for the macrofunction using control register(s): operation mode registers or configuration registers. Each bit in this register(s) is associated with the

TABLE 3.1

Types of on-chip PEs and their functional configuration/programming

Type of PE in the FCR	Fine-grained PE in the FPGA	Fine-grained PE in the CPLD	Coarse-grained PE in the FPGA	Hybrid PE in the FPGA
Method of functional programming	Loading data to the LUT	Programming status of pass transistors	Loading data to the control/ op-mode registers	Loading program to the program memory
Best class of application	Computation-intensive tasks	Fast glue-logic and custom interfacing	Application-specific (e.g., DSP) computation-intensive tasks	Algorithmic-intensive tasks combined with computation-intensive tasks

control input of the mode-switching multiplexer. The set of control register bits allows rapid and flexible determination of the macro-operation in the coarse-grained PE. The advantage of the coarse-grained PE is the ability for parallel data-processing in spatial and temporal domains: (1) processing multibit words instead of single-bit operations in the fine-grained PE and (2) pipelined data-stream execution due to compact allocation of all PEs (e.g., add/subtract units, dedicated multipliers) and pre-/postoperative registers. Thus, coarse-grained PE is most effective for application-specific computation-intensive macro-operations (e.g., FFT, FIR, FP arithmetic).

At the same time, the control dominant algorithmic-intensive operations may not be suitable for the above coarse-grained PEs. Therefore, for applications that contain a large portion of algorithmic-intensive operations and do not have strict performance constraints, the PEs based on CPU-centric instruction processors may be a more effective solution. The algorithm programming for this type of PEs is conducted using traditional memory modules—usually composed of instruction and data-cache memory modules L_1 and often L_2 levels.

Summarizing the aforementioned, Table 3.1 describes the types of the on-chip PEs in the FCR and the method of configuration/programming of the required functionalities.

Exercises and Problems

Exercises

1. Define two levels (stages) of the on-chip architecture configuration in the FCR.
2. List three possible mechanisms for configuration and programming of the on-chip PEs.

3. What are the two ways to get the results of functional operations in a processing element?
4. Define the pros and cons of implementation of the fine-grained processing elements based on LUT in comparison to implementation of logic gates.
5. Determine the reasons that cause timing uncertainties and output signal oscillation in logic circuits based on (a) LUT and (b) PLA.
6. How can the LUT/PLA output be stabilized eliminating output signal oscillation?
7. Define the method for clock period determination for LUT-/PAL-based logic circuits.
8. What is the main reason to combing multiple LUTs and DFFs in one CLB in modern FPGA devices?
9. Why is the clock frequency and performance of the dedicated hardware circuits deployed in the ASIC always higher than the same hardware circuits built on the base of reconfigurable CLBs or PLAs?
10. What are the pros and cons of the coarse-grained FPUs in comparison to fine-grained configurable processing elements? Can coarse-grained FPUs provide higher performance than fine-grained PEs? If yes, what are the main conditions?
11. What benefits can reconfigurable ALU give in the coarse-grained FPU and why?
12. What is the main reason for hybrid architecture organization in RCS? When is this organization effective and when is it not needed?
13. List the reasons and benefits for on-chip embedded PPEs.
14. What are potential drawbacks of hybrid architecture with embedded hard-core PPEs?

References

1. R. L. Tokheim, *Digital Electronics: Principles and Applications*, 6th ed., Glencoe/McGraw-Hill, Columbus, OH, 2003, Chapter 4.
2. S. Brown and Z. Vranesic, *Fundamentals of Digital Logic with VHDL Design*, 2nd ed., McGraw-Hill, New York, 2005, Chapter 3.
3. M. A. Miller, *Digital Devices and Systems with PLD Applications*, Delmar Publishers, Albany, NY, 1997, Chapter 6.
4. Xilinx Inc., *7 Series FPGAs Configurable Logic Block*, User guide, UG474 (v1.7), 2014, Chapter 2.
5. Xilinx Inc., Understanding the CoolRunner-II logic engine, Application note CoolRunner-II CPLD, XAPP376 (v1.0), 2002.
6. Xilinx Inc., *7 Series DSP48E1 Slice*, User guide, UG479 (v1.6), 2013, Chapter 2.
7. A.V. Oppenheim, R.W. Schafer, and J. R. Buck, *Discrete-Time Signal Processing*, 2nd ed., Prentice Hall, Upper Saddle River, NJ, 1999.

8. V. Baumgarte, F. May, A. Nuckel, M. Vorbach, and M. Weinhardt, Pact-XPP—A self-reconfigurable data processing architecture, *in Proceedings of the International Conference on Engineering of Reconfigurable Systems and Algorithms, CSREA-2001,* CSREA Press, Las Vegas, NV, June 2001, pp. 64–70.

9. M. Gokhale and P. S. Graham, *Reconfigurable Computing: Accelerating Computation with Field Programmable Gate Arrays,* Springer, Dordrecht, the Netherlands, 2005, Chapter 2.

10. Altera, Cyclone V hard processor system technical reference manual, CV_5v4, Altera complete design suite: 15.0, May 2015.

11. Xilinx Inc., *Embedded Processor Block in Virtex-5 FPGAs,* Reference guide, UG200 (v1.8), February 2010.

12. Xilinx Inc., *Zynq-7000: All Programmable SoC,* Technical reference manual, UG585 (v1.10), February 2015.

13. Microsemi, *SmartFusion Microcontroller Subsystem,* User guide, MSS_UG V4: 50200250-4/07.14.

14. J. W. Valvano, *Embedded Systems Introduction to Arm® Cortex(TM)-M Micro-Controllers,* 5th ed., Vol. 1, CreateSpace Independent Publishing Platform, Middletown, DE, 2015.

4

Reconfigurable Communication Infrastructure in the FCR

4.1 Introduction

This chapter is dedicated to the aspect of configuration and reconfiguration of communication resources at the on-chip level in the field of configurable resources (FCR) as well as interfacing the on-chip communication network to the system level of reconfigurable computing systems (RCS). In Chapter 3, configuration of functional components on the base of uniformed fine-grained, coarse-grained, and hybrid processing elements (PEs) was discussed. However, according to the definition of computing architecture given in Chapter 1, the architecture consists of functional components, links between these components, and procedures for data acquisition, execution, storage, etc. In contrast to conventional computing systems, RCS architecture allows configuration and reconfiguration of the topology of communication links between the functional components—PEs. This allows static and dynamic adaptation of the RCS architecture to the task algorithm and data structure as well as performance constraints. Therefore, after the discussion of PE configuration and reconfiguration provided in Chapter 3, it is necessary to consider the mechanisms for communication link configuration including (1) the creation of links between PEs at the on-chip level of the FCR and (2) the configuration of input/output (I/O) interfaces for information exchange between on-chip and on-board level of the FCR. Due to the large variety of approaches for establishing the configurable communication links between PEs and configurable interfaces to peripherals in the RCS, it is difficult to observe all of them in this book. Therefore, only the main approaches practically implemented in the FCR at the on-chip level in modern FPGAs and complex programmable logic device (CPLDs) will be considered in this chapter.

4.2 Organization of the On-Chip Communication Infrastructure

As mentioned earlier, there are three levels of RCS architecture: on-chip, on-board, and multiboard system levels. It is necessary to mention that system level could be limited by on-chip only (system on programmable chip [SoPC]). Nonetheless, the on-board level of the system exists in most practical cases. However, consideration of the RCS as a multiboard system is also possible. In this case, system level of the FCR should be extended to multiple boards. However, the on-chip level of architecture provides the highest flexibility in the organization of the data communication infrastructure because of the following reasons:

- Number of communication lines and switching elements that could be deployed on a die is much higher than the same resources allocated to on-board or multiboard level.

- Distances between components and, thus, distances of communication lines at the on-chip level are shorter in orders of magnitude than in on-board or multiboard levels.

- The pinout of the on-board components as well as layout constraints of on-board signal layers are strong limiting factors for the communication bandwidth, power consumption, and reliability of the on-board communication network. All these constraints are even stronger for the multiboard level of system organization.

Thus, the designers are trying to deploy communication and switching elements at the on-chip level of the FCR as much as possible. These elements must provide communication links, shared buses, and I/O interfaces between the following components allocated at the on-chip level:

1. PEs configured or programmed for given functions for the period of task execution
2. On-chip data storage elements (e.g., SRAM modules or block random access memory [RAM] units)
3. Configurable on-chip interface elements such as I/O blocks, which can provide defined electrical standards for on-board level communication (e.g., low-voltage transistor-transistor logic [LVTTL], low-voltage differential signaling [LVDS])
4. Multigigabit serializers/deserializers (SerDes), which can provide high-bandwidth information exchange at the on-board and multiboard level of communication

5. Bus interfaces, which can be configured for the required standard of communication (e.g., Peripheral Component Interface [PCI] bus or double data rate [DDR]-synchronous dynamic random access memory [SDRAM] memory bus)

The concept of the FCR organization presented in Chapter 1 (Figure 1.2) assumes that identical uniformed data-processing, data storage, and I/O interface elements are allocated between routing elements. These routing elements consist of routing lines and programmable switching elements (PSEs). Thus, routing elements could be programmed to configure almost any topology of interconnects between the aforementioned processing, storage, and interfacing elements in the FCR. The 2D nature of the die allows only 1D (bus or pipeline) or 2D (matrix) organization of the routing topology. Thus, routing elements should provide 1D (bus) connection to/from PEs and/or 2D (vertical/horizontal) interconnects to allow PE information exchange via 2D communication topology. In all cases, the routing element consists of (1) routing lines (wires) and (2) *programmable switch* that may connect these lines together or disconnect them. The simple programmable switch can be implemented as the pass transistor with the gate connected to the output of 1-bit configuration memory cell. This element is described in Section 3.2 (Figure 3.5).

4.3 Fine-Grained On-Chip Routing Elements in the FCR

The programmable switch is used everywhere in the on-chip interconnect points where two crossing wires may or may not be connected. In other words, this programmable switch can be considered as the simplest "fine-grained" 1-bit PSE. However, routing resources can be utilized more effectively if cross point is organized as "multidirectional." This organization of interconnect point based on six pass transistors [1] is shown in Figure 4.1.

In contrast to one-pass-transistor programmable cross switch depicted in Figure 3.5, the six-pass-transistor-based interconnect point can provide up to 64 combinations of interconnects between four communication wires coming to this cross point: *LN*, *LE*, *LS*, and *LW*. For example, if switches *S1* and *S6* are closed, the "west" line *LW* transmits signal to the "north" line *LN*, and the "east" line *LE* is connected to the "south" line *LS*. All other switches, *S2*, *S3*, *S4*, and *S5*, are open in this case. Horizontal lines (*LE* and *LW*) and vertical lines (*LN* and *LS*) can be interconnected at the cross point by closing switches *S1*, *S2*, and *S3* when the rest of the switches are open. If the interconnection in this cross point should be bypassed, only switches *S3* and *S4* should stay connected.

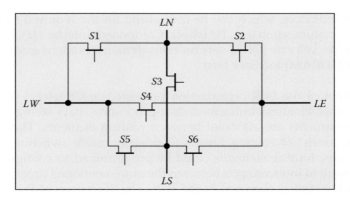

FIGURE 4.1
Multidirectional six-pass-transistors interconnect point.

Therefore, this six-transistor-based interconnect point provides much higher flexibility for the "fine-grained" (1 bit/line) interconnect point in comparison to one-pass-transistor-based programmable switch. On the other hand, programming of this interconnection point requires 6 bits in the configuration memory because each gate of the pass transistor has to be connected to the respective memory cell, as shown in Figure 4.2.

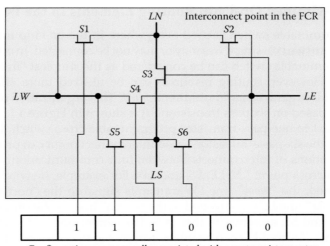

FIGURE 4.2
Programming the interconnection pattern to the interconnect point in the field of configurable resources.

4.4 Coarse-Grained On-Chip Routing Elements in the FCR

As per organization of configurable PEs discussed in Chapter 3, all types of PEs require multibit parallel inputs and for coarse-grained PEs, multibit (word-wide) outputs. Therefore, PSEs should be combined to the multibit units to transfer data words. In contrast to application specific integrated circuits (ASICs) or microprocessors, the length of data words in the FCR could vary in a certain range. In other words, different PEs may have different width of input and output data words. Hence, flexibility in creation of communication links with custom data word length became one of the most important aspects in the organization of the FCR.

One of the useful on-chip multibit routing units is the programmable switch matrix (PSM) [2]. This unit combines the array of the aforementioned six-pass-transistors programmable switches. The organization of this PSM is depicted in Figure 4.3.

The six-pass-transistors-based interconnect points are allocated in the diagonal of the PSM providing very flexible configuration of multibit links between PEs (e.g., CLBs in FPGAs).

On the other hand, different PEs may require different configurations of their inputs and outputs.

That is why certain composition of 1-bit cross point programmable switches with short routing lines and PSM combined for long routing lines was determined as an effective solution [2]. The diagram of such composition is shown in Figure 4.4.

It is necessary to mention that on-chip organization of interconnects between PEs in the FCR may utilize PSM and PSEs. This combination can provide more flexibility in PE interconnection by using short communication lines for local interconnects between neighboring PEs and PSMs and

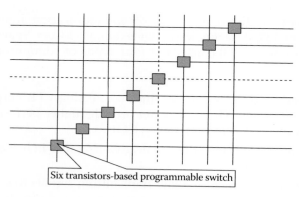

Six transistors-based programmable switch

FIGURE 4.3

Organization of the programmable switch matrix in the on-chip field of configurable resources.

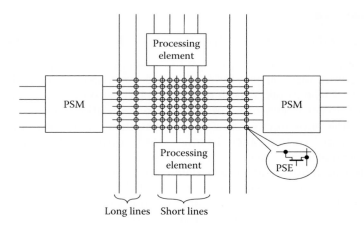

FIGURE 4.4
Combined PE interconnects using programmable switch matrix and programmable switching elements.

long lines for data exchange between the PEs allocated relatively far from each other. The long lines play the role of highways in comparison to city streets. These long lines do not have many traffic lights as in the streets and therefore do not reduce the average speed of cars. Taking this analogy, it is possible to say that long lines do not have as many switching elements between communication nodes as short lines. Each pass transistor, however, adds its capacitance to the communication line and may change the direction of signal transfer. Thus, the complete impedance of the communication line is not static and depends on the number of switching points active at a time. The impedance of the line influences on signal propagation time between nodes. Thus, it influences on the speed of communication and finally on system performance. Hence, utilization of long lines for communication allows performance acceleration.

This concept of different length of communication lines is widely used in recent FPGA technology. For example, Xilinx Spartan-6 FPGA routing infrastructure [3] consists of four types of communication lines, as follows:

1. "Fast interconnects"—type of lines connecting CLB outputs with other CLB inputs. These lines can provide higher performance for simple functions deployed only in CLBs.

2. "Single interconnects"—lines to neighboring tiles in horizontal and vertical directions.

3. "Double interconnects"—allow transmitting signals to every other tile in the FPGA device in horizontal and vertical directions and to the diagonally adjacent tiles as well.

4. "Quad interconnects"—connect to one out of every four tiles in horizontal, vertical, and diagonal directions. It was determined that this
method can provide higher flexibility than the long lines that were
used in previous generations of Xilinx FPGAs.

Thus, the on-chip communication infrastructure oriented on data-transfer
between different PEs consist of both (1) fine-grained routing elements
including PSEs and (2) coarse-grained PSMs coupled with communication
lines with different length and directions: (1) short and direct; (2) double
and quad length to vertical, horizontal, and diagonal directions; and (3) long
lines to vertical and horizontal directions.

This organization allows the most flexible mapping of the task segments
on PEs.

However, there could be certain reasons when the FCR could not be allocated only at the on-chip level. In this case, the FCR should be extended
on the on-board or multiboard (system) level of design. This aspect of the
FCR organization will be discussed in Chapter 5. In all cases, the on-chip
level of FCR must be interfaced with circuits deployed on the system level
of RCS architecture. Therefore, all existing programmable logic devices
(PLDs) are equipped with respective I/O resources for data exchange
between on-chip and off-chip levels of the RCS.

4.5 Fine-Grained On-Chip Configurable
Input/Output Elements

The fine-grained I/O element (IOE) should provide digital synchronous
receiving (from printed circuit board [PCB] to system-on-chip [SoC]) or
transmitting (from SoC to PCB) of 1-bit information at a time.

Since the electrical standards for signal transmission in the on-board environment are different than the on-chip standards, certain adaptors as electrical circuits must be a part of the IOE.

On the other hand, the impedances of the on-chip routing lines are very
different from the on-board signal transfer lines. Therefore, the power
required for the on-board signal transmission can be in the order(s) of magnitude higher than for the on-chip signal transfer. Hence, the IOE should
have circuits for digital control of impedances as well as power conversion
circuits.

Another aspect associated with the on-chip routing lines is the relatively
low cost and large number of these lines in the on-chip infrastructure of
PLD. This allows reaching high bandwidth in data-transfer by parallel data
transmission between PEs, embedded memory blocks, IOE, etc.

In contrast to the on-chip data transmission, PCB is unable to provide as many routing lines as the on-chip infrastructure. The distances for data-transfer are in orders of magnitude longer than the on-chip ones. Therefore, it is necessary to utilize special techniques to increase the bandwidth of on-board lines. The differential signaling and *DDR* data-transfer [4] became the main mechanisms for better utilization of a limited number of the on-board lines.

Additionally, the IOE should be able to interface on-chip circuits to the on-board shared buses. Thus, it should provide tri-state output to avoid multi-sourcing [5]. All the aforementioned features may or may not be required for a particular interface in a particular application. Hence, the IOE circuit must be configurable prior to the operational mode and reconfigurable during the mode of operation.

Accordingly, the generic organization of the IOE should contain the following circuits:

1. The input buffer (IBUF) able to receive 1-bit data from the on-board communication line according to selected electrical and signaling standard
2. The tri-stated output buffer (OBUF) for transmission of 1-bit data to the number of on-board receivers via PCB line according to selected electrical and signaling standard
3. The input and output registers (flip-flop–based) to latch the 1-bit data
4. The line impedance control circuit(s) able to match selected type of line termination according to the distance, line impedance, and system clock frequency requirements

In addition, the circuit providing the DDR transfer can be included in the IOE for better utilization of on-board communication lines specialized for DDR. It is necessary to mention that particular families of FPGA devices may have many additional features and support different industrial communication standards. It is difficult to observe all of them in this general overview. Thus, only those basic features of associated circuits will be discussed next.

4.6 Generic Organization of Input and Output Buffers for PCB Interface

The goal of IBUF is to receive the digital signal from the communication line and convert it to the level of logic 0 or 1 acceptable for on-chip utilization. Usually, complementary metal-oxide semiconductor (CMOS) circuits are

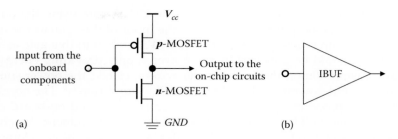

FIGURE 4.5
Complementary metal-oxide semiconductor gate–based input buffer (IBUF) (a) and IBUF symbol (b).

used for this purpose. An example of CMOS IBUF is shown in Figure 4.5a and its symbol is depicted in Figure 4.5b.

Here, the input signal comes from the on-board components and the output generates the respective signal to the on-chip circuits (e.g., inside the FPGA device). If the input voltage is in the range of "low" level (close to 0 V), the *p*-MOSFET transistor behaves as the closed switch and, thus, connects the output of this gate to the V_{cc}, the voltage source for the on-chip circuits.

Alternatively, if the input voltage is in the range of "high" level, the *n*-MOSFET connects the output to the ground when the *p*-MOSFET is open. Thus, the IBUF could be considered as the single pole double throw (SPDT) switch controlled by the input voltage level as shown in Figure 4.6. Indeed, the output of this buffer can be connected to the voltage source of on-chip circuits, V_{cc}, or ground, *GND*, depending on the input state "0" or "1." For example, if the IBUF is based on noninverting CMOS gate, the $V_{out} = GND = 0$ V when input = logic '0" and $V_{out} = V_{cc}$ when input = logic "1."

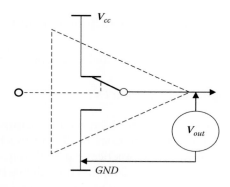

FIGURE 4.6
SPDT model of the complementary metal-oxide semiconductor gate–based input buffer.

The OBUF can also be based on the same CMOS gate using the output of the gate as the output of the chip and the input of this gate as the input from the on-chip routing lines. In this case, V_{cc} applied to this gate as voltage source should be powered by the power supply providing energy for signal transmission on the board or the system. Hence, the power sources for on-chip and system levels of the same PLD are different. The necessity for different power supplies for the on-chip circuitry and on-board communication network is dictated by different power requirements for circuit operation inside the chip and signal transfer between the chips on long distances on the PCB. Indeed, the power source for the on-chip data transmission may have lower voltage than the power supplies for the on-board communication lines because short distances of on-chip routing lines allow reaching sufficient noise immunity with relatively low core voltage level. On the other hand, low core voltage, $V_{cc(on-chip)}$, minimizes the entire power consumption of the on-chip resources. Thus, most of the FPGA devices and CPLDs require different power sources for the FCR deployed in the device and I/O interface circuits.

Let us consider the data-transfer process from the OBUF in on-board component A to the IBUF located in component B deployed on the same board. Figure 4.7 illustrates this process.

The on-board data transmission line has certain impedance, $Z = f(R, L, C, G, Fs)$, which is the function of resistance R, inductance L, capacitance C, conductance G, and signal frequency Fs. In the general model of telegrapher's equation of the transmission line [6], the impedance of the transmission line is $Z = \sqrt{R + j2\pi Fs * L / G + j2\pi Fs * C}$. As for this equation, the data transmission line impedance depends on

1. The resistance R of OBUF and IBUF transistors, chip pads, and transmission line
2. Composite inductance L of the pads and topology of the transmission line
3. Aggregated capacitance C of OBUF and IBUF CMOS gates and transmission line
4. Conductance G associated with dielectric characteristics of signal layers
5. Frequency of signal transmission Fs

For relatively low frequencies of on-board data transmission (<1 MHz), the aforementioned impedance usually does not bring much influence on the data-transfer process. However, in practically useful frequencies in the range from hundreds of MHz to units of GHz, this impedance plays a crucial role. That is why different termination schemes may be needed for proper layout of each on-board communication line. Thus, the device with reconfigurable

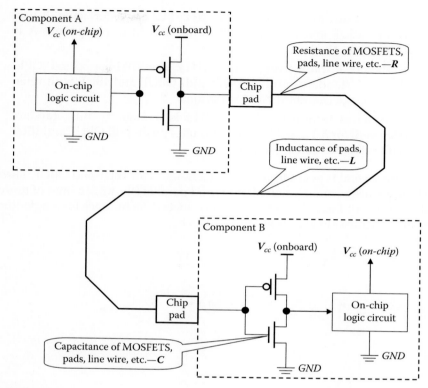

FIGURE 4.7
Connection of component A and component B via data transmission line.

architecture (e.g., FPGA) should be equipped with configurable I/O impedance circuits to maintain signal integrity (SI) [7]. Otherwise, extension
of the FCR to multichip implementation will become much more difficult
or even impossible. Same is true for interfacing the reconfigurable device
with high data rate memories, sensors, and actuators requiring high-speed
data-transfer. To avoid signal reflections, ringing in the on-board data transmission lines, several I/O standards, and associated line (trace) termination techniques are utilized in most of FPGA devices for single-line-per-bit
transmission:

1. LVTTL: the EIA/JESDSA standard single line transmission general
   purpose I/O (GPIO) using I/O source voltage—$V_{cc}=3.3$ V.
2. Low-voltage CMOS (LVCMOS) for different voltages: 2.5 V for
   LVCMOS25, 1.8 V for LVCMOS18, and 1.5 V for LVCMOS15. All these
   standards also are single line transmission standards for GPIO
   interfaces.

3. PCI standard for 33, 66, and 133 MHz PCI bus interfacing. It uses LVTTL IBUF and push–pull OBUF (as shown in component A in Figure 4.3). The I/O $V_{cc} = 3.3$ V.

4. High-speed transceiver logic (HSTL): the IBM-supported GPIO (JESD8-6 standard) utilizes push–pull OBUF and IBUF with reference voltage (standard has 4 variations).

5. Stub series terminated logic (SSTL) is a memory bus standard (JESD8-8) for 3.3 and 2.5 V that also uses push–pull OBUF and IBUF with reference voltage source.

There are several types of termination techniques for IBUF and OBUF. These techniques are a combination of series and parallel (shunt) schemes of termination resistors. Generic examples of termination techniques for single-line-per-bit transmission are shown in Figure 4.8.

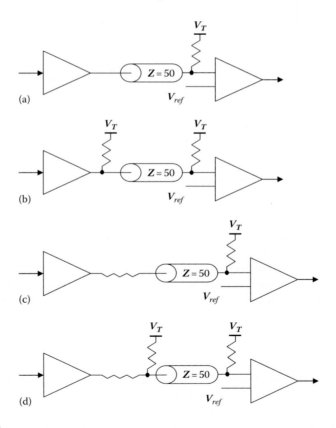

FIGURE 4.8
Examples of termination techniques for single-line-per-bit transmission lines. (a) Parallel terminated input, (b) parallel terminated input and output, (c) series terminated output and parallel terminated input, and (d) series-parallel terminated output and parallel terminated input.

HSTL-IV DCI transmitter and receiver termination

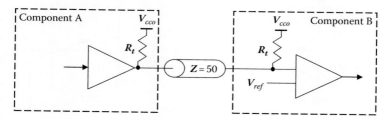

FIGURE 4.9
Example of termination of the high-speed transceiver logic class IV communication line.

The actual method of line termination is usually specified by the standard of communication and in the data sheet of the selected PLD (e.g., Xilinx Spartan [8] or ALTERA MAX [9]).

For example, the required termination of the HSTL-IV class communication is shown in Figure 4.9. The termination resistors R_t are embedded in the device and could be connected to the OBUF output as well as to the input of the IBUF by configuration code (certain bits in the configuration bit stream). The mechanism for impedance on-chip control is called digital controlled impedance (DCI) [10]. This mechanism allows configuration of interface standards and associated termination techniques by programming attributes for IOEs in the PLD.

4.7 Generic Organization of Output Buffers for Dynamic Links (Buses)

In contrast to statically configured on-board links between different PLDs, there may be a need for dynamic links between components. We can consider a link as dynamic when communication between two components is not needed for the entirety of the time of task execution. In this case, the shared bus can be an effective solution. In the shared bus, each line for data-transfer from the $OBUF_i$ of one component C_i to the $IBUF_j$ of another component C_j may be assigned for some period of time. After the period of transaction between these components, the next couple of components (e.g., C_k and C_s) may use the same bus line for communication. Alternation of outputs of components C_i and C_k connected to the same wire requires disconnection of the output of nonactive component from this wire. Otherwise, the so-called *multisourcing* situation may occur on this line. The multisourcing can cause physical damage to one of the OBUF transistors and permanent hardware fault in the associated PLD. To illustrate this effect, let us

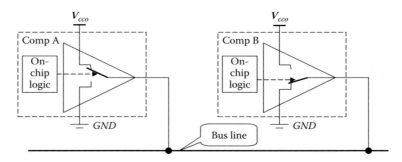

FIGURE 4.10
Example of multisourcing problem occurring in the shared bus line.

consider the example of two push–pull OBUFs connected to the same bus line. This circuit is shown in Figure 4.10.

If the OBUF of component A sends the bit equal to "1," the metal-oxide semiconductor field effect transistor (MOSFET) switch is connected to the positive rail of V_{cco}. At the same time, the OBUF output of the nonactive component B is set to "0." It means that the MOSFET switch of this component is connected to the ground (GND). Hence, the bus wire creates the short circuit because the V_{cco} rail is connected to the ground rail via the MOSFET switches of component A and component B. As a result of this *multisourcing*, one of the transistors in OBUF A or OBUF B will work as a fuse and, thus, may burn down.

To avoid the multisourcing problem, each OBUF must have the pass transistor (pass switch) to tri-state the buffer output (disconnect from the shared data transmission line) as shown in Figure 4.11a. The figure shows that tri-state circuit output is the output that can be in one of the three possible states:

1. Connected to positive voltage rail providing *low*-output impedance
2. Connected to the ground rail providing *low*-output impedance
3. Disconnected from the ground or power rails providing *high*-output impedance

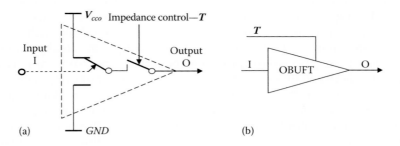

FIGURE 4.11
(a) Tri-state output buffer organization and (b) symbol.

Thus, the tri-state OBUF (T-OBUF or OBUFT) must have the additional control input (T) to manage the output impedance of the buffer. The control bit associated with this impedance control circuit is not included in the configuration bit stream but is part of the on-chip component architecture. This organization allows to connect the component's OBUF to the bus line only for the periods when data-transfer is needed and disconnect at the end of data transaction. The *symbol* of OBUFT is shown in Figure 4.11b.

Let us consider the same example, as depicted in Figure 4.10, in the case when OBUFT is used. This case is shown in Figure 4.12. In this example, the output pass switch is closed in the OBUFT of component A, because control $T=1$. Thus, the output voltage close to V_{cco} is applied to the bus line. At the same time, the output pass switch of component B is open, and therefore, high impedance of this output does not influence on the voltage level of the bus line. Hence, if only one component's output is connected to the bus line at any moment of time, the multisourcing will never occur. Let us consider multiple IBUFs attached to the bus line as the receivers of data. Due to the structure of CMOS gates (e.g., Figure 4.5), their input impedance is relatively high and thus do not influence much on the voltage level of the associated bus line.

Therefore, the generic I/O buffer (IOB) is a composition of the IBUF and OBUFT combined with the I/O data registers (flip-flops), tri-state control register (for *T*-bit), and DCI for I/O impedance configuration. This IOB structure is shown in Figure 4.13.

Initially, the pad of the IOB is set up to high impedance and ready to receive data from the data-transfer line. When data is coming to the pad of the on-chip logic circuits, it will be registered in the input data register according to the clock signal from the on-chip clock generator. In the case when data should be transmitted from the IOB, it should be written first to the output data register. When the shared communication line is free from other data sources, the respective signal comes to the tri-state control register and, thus, connects the OBUFT output with the communication line via pad.

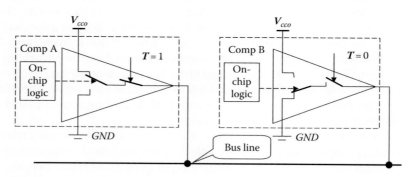

FIGURE 4.12
Avoiding multisourcing problem using tri-state buffers.

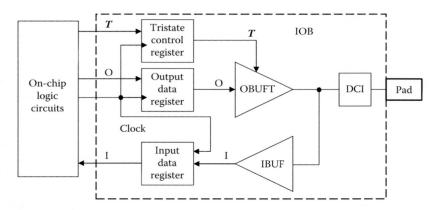

FIGURE 4.13
Generic structure of the input/output buffers.

4.8 Reduction of Electromagnetic Noise and Crosstalk in Data-Transfer Lines

Since the RCS is focused on high-performance computing, the on-board communication infrastructure should support high-bandwidth data-transfer. This, in turn, requires transmission of signals on PCB with high clock frequencies. These frequencies are usually in the range of hundreds of MHz to units of GHz. The signals being transmitted in VHF class of frequencies can generate relatively high electrical noise that can influence other signal wires or even the entire devices sensitive to RF noise. As a result, the system may suffer from communication failures, data-transfer errors, etc., caused by electromagnetic noise. The level of the aforementioned noise and associated crosstalk depends on (1) peak-to-peak voltage of signals, (2) length of the transmission lines, and (3) frequencies of signals. In other words, these transmission lines can be considered as a sort of RF antennas allocated on the board. In this light, OBUFs can be considered as RF transmitters. On the other hand, IBUFs connected to communication lines are working as RF receivers. Furthermore, the transmitting lines (antennas) and receiving lines often are allocated in a range of millimeters from each other. Therefore, crosstalk between signal lines may be very high and, thus, unacceptable for communication between on-board devices.

Therefore, for moderate frequencies of data transmission (in the range of 10–66 MHz), single-ended lines are effective for on-board communication for distances <0.5 m. In this case, single-ended lines give certain benefits for the board layout because they require only one pad of the device and one transmission line per bit of data-transfer. As per [11], CMOS/TTL standards can provide reliable data-transfer rates at 33 MB/s for distances (on-board) up to 500 mm or less. The SSTL standard may pull up this data rate up to 200 MB/s for the same range of distances [12].

However, for higher data rates (in the range of 100 MB/s to 1 GB/s) for distances up to 1 m, the *differential signaling* is often used instead of single-ended signaling standards. The concept of differential signaling assumes utilization of two wires instead of one transmission line. In addition to that, the *differential push–pull driver* must be used as a transmitter. Also, the differential signal receiver (op-amp based) should be used on the receiving end. The generic organization of a point-to-point link based on differential signaling is shown in Figure 4.14.

There are two most common standards associated with data exchange using differential signaling concept: LVDS, specified in the standard TIA/EIA-644 (introduced in 1996), and multipoint LVDS (M-LVDS), specified in standard TIA/EIA-899 (introduced in 2002). The details of implementation of LVDS data-transfer can be found in many literature sources (e.g., Reference 13). The major benefits of LVDS/M-LVDS data-transfer in on-board communication infrastructure are the following:

- High signal–noise ratio is provided by differential signaling because the differential receiver operational amplifier (Op-Amp) subtracts voltage induced by noise on one line from the same level of noise voltage induced on the other wire at any moment of time. This is correct in the case when both signal lines of each differential pair are equal to each other in distance, topology, impedance, etc. Therefore, for differential signaling, the *distance matching* and *impedance matching* for each line in the pair are a must.

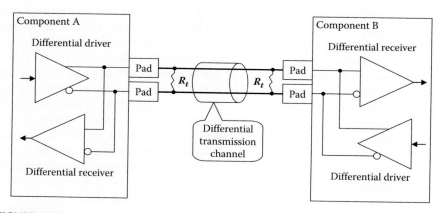

FIGURE 4.14
Generic organization of a point-to-point differential data interchange link.

- Low level of noise generation by LVDS lines is insured by low level of peak-to-peak differential voltages in the pair of wires. This is possible because the complementary current in each wire of the pair cancels each other's generated electromagnetic fields. Therefore, the differential bus combined by closely allocated parallel differential pairs of lines is more resistant to crosstalk than the bus based on single-ended lines.
- Bandwidth of differential pair of lines could be in the order of magnitude higher than the same single-ended line. At the same time, the distance for reliable data transmission is 2–3 times longer than for comparable single-ended line.
- The power consumption per bit transfer is lower than in single-ended transmission due to smaller voltage span in differential signaling compared to single ended.

Therefore, the IOBs for the fine-grained on-chip configurable IOEs should be able to configure the IBUF and OBUF as single-ended bit-transfer elements or as differential depending on distance and required bandwidth of data transmission. In case of short distances and relatively low data rates, the single-ended option may be optimal because it allows the use of only one signal line per bit. Alternatively, in case of high bandwidth and/or longer distances required for data-transfer, the differential signaling standards may be determined as optimal. That is why most of the vendors of FPGA devices include the aforementioned features in their devices.

4.9 Increasing Bandwidth Using DDR Transmission

As mentioned earlier, the major constraint for the on-board communication infrastructure is the limited number of communication lines and associated I/O interface blocks in devices deployed on-board. On the other hand, on-chip data-transfer rate is much higher than available bandwidth of data-transfer lines in the on-board level of the RCS. Since the clock frequency for signal layers are limited by impedances and distances of transmission lines, the most effective way for data rate increase is utilization of the concept of double data rate (*DDR*). This concept assumes synchronization of data words at the rising edge of the clock and the falling edge of clock signal supporting respective data bus. In other words, I/O interface of a sender generates a pair of data words per one clock cycle. The first word of this pair is synchronized by the rising edge of the clock signal and the second word is synchronized by the falling edge, as shown in Figure 4.15. This method allows doubling of the data-transfer rate using the same clock frequency as it was used for the regular (word-per-clock cycle) method of data transmission.

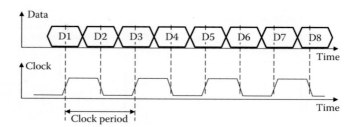

FIGURE 4.15
Data-transfer utilizing double data rate method of transmission.

This figure shows that during one clock cycle period, the data bus is sampled twice: (1) at the moment of rising edge of the clock signal and (2) at the moment of falling edge of the same clock.

Thus, the receiver's register reg-1 is writing the input data D1, D3, D5, and D7 at the rising edge of the clock, and the other register reg-2 is writing data D2, D4, D6, and D8 being synchronized by the falling edge of the clock (rising edge of inverted clock signal). Two registers are necessary in this case because D-flip-flop-based registers can write data only in one edge of the clock signal.

The organization of 1-bit DDR receiver circuit is depicted in Figure 4.16.

The transmitting circuit providing DDR data-transfer is shown in Figure 4.17, and the timing diagram of the DDR transmitting process is depicted in Figure 4.18.

As seen in Figures 4.17 and 4.18, two data sources are providing the output data according to common clock frequency. The transmitting data-stream can have DDR by using the DDR transmitting circuit. This circuit consists of two registers, reg-1 and reg-2, synchronized by the rising edge of Clk and the falling edge of Clk, respectively. The circuit output is the output of the multiplexor MUX. The data (bit equal 0 or 1) appears at the output of data

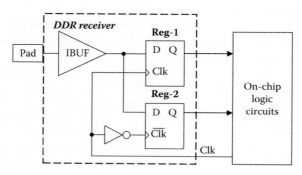

FIGURE 4.16
Organization of the 1-bit double data rate receiver.

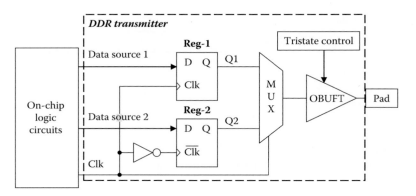

FIGURE 4.17
Organization of the 1-bit double data rate transmitter.

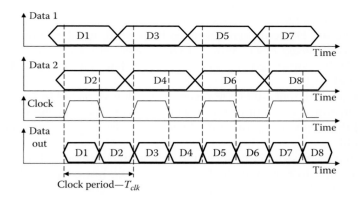

FIGURE 4.18
The timing diagram of double data rate transmitting process.

source 1 and data source 2 almost simultaneously. However, some delays between *data-1* and *data-2* could appear due to propagation delays in the on-chip routing lines and switches. Nonetheless, the period of *data-1* and *data-2* is equal to the common clock period T_{clk}.

It is assumed, however, that on-chip logic circuits provide *data-1* valid at the moment of the rising edge of the clock signal, and *data-2* is valid when the falling edge of the clock (rising edge of inverted clock) signal occurs. At the same time, multiplexor MUX switches to the reg-1 output when clock signal reaches "1" and to the reg-2 output when clock is "0." In the first half of the clock cycle period, the output of buffer OBUFT transmits *data-1* (e.g., D1, D3, D5, D7), and in the second half of the clock period, it transmits *data-2* (e.g., D2, D4, D6, D8). Therefore, during one clock cycle, both data-elements (from data sources 1 and 2) are transmitted.

Practically, the DDR method of transmission often is combined with differential transmission lines. The reason is that DDR data-transfer usually requires twice higher clock frequency than the on-chip common clock frequency, and therefore, this frequency is in the range of hundreds of MHz and may even reach the GHz range. That is why the LVDS or similar standards of transmission are used in case of DDR data transmission. Usually, this method is implemented in memory interfaces such as DDR-SDRAM memory modules but can be used for high-speed full-duplex data transmission between FPGAs or other processing units supporting DDR standards.

Actually, the DDR transmission could be considered as "compensation" for utilization of a pair of transmission lines in differential data-transfer channel. Certainly, the additional register and multiplexor is needed for that. Therefore, additional hardware in DDR circuits and additional communication elements in differential communication channel (dual line, op-amp, differential driver, etc.) allow significant increase in the data rate and distance for communication as well.

4.10 Coarse-Grained Interface Elements for On-Board Communication

In most practical application links between on-board processing units, memory elements and interfaces to peripheral devices should provide more than 1 bit/transaction. Thus, certain combination of fine-grained IOEs is necessary to provide parallel data-transfer. Therefore, fine-grained IOEs allow designing custom point-to-point links or buses with programmable width, data rate, and impedance. This is a very valuable feature that makes static configuration of data-transfer highways optimized to the requirements of an application possible, and these custom word-wide communication links can be considered as a coarse-grained interface elements.

Another type of coarse-grained interfaces built-in to the reconfigurable components (e.g., FPGAs) is application-specific modules providing communication via standard buses or bus interfaces. In this case, vendors include the embedded coarse-grained interface modules, such as PCI or PCIe interface modules (e.g., Xilinx Zynq-7000 series FPGA [14]), or interfaces to DDR-SDRAM memory modules (e.g., Xilinx Virtex FPGAs [15] or ALTERA Stratix FPGA [16]).

On the other hand, on-chip PE can generate the output results in the range of hundred(s) of megabytes per second when the DDR LVDS communication line can provide data rate in the range of gigabits per second. There are two possible solutions for a designer of the on-board communication infrastructure to balance the PE output rate and data transmission bandwidth: (1) to use the parallel bus working with relatively low frequency or (2) to implement serial bus with relatively high frequency.

In case of parallel bus, the LVTTL single-ended lines and associated IOEs can be combined to the custom bus (e.g., 16 or 32 lines/bus). For example, in case of 100 MB/s output data rate, the bus with 16 parallel single-ended lines will require frequency computed as follows:

$$F_{bus} \text{ (16-bit parallel)} = 100 \times 1024 \times 1024 \times 8 \text{ bits/16 bit per clock cycle}$$
$$= 52.429 \text{ MHz}$$

This bus can be implemented using LVTTL single-ended lines with relatively low power consumption but wider PCB area utilized for this bus. The alternative could be serial transmission via LVDS bus with two differential lines. For the aforementioned case, bus frequency can be calculated as follows: F_{bus} (serial) $= 100 \times 1024 \times 1024 \times 8$ bits/s $= 838.860$ MHz.

If the DDR data transmission is implemented, the bus frequency will be twice lower:

$$F_{bus} \text{ (serial-DDR)} = 838.860 \text{ MHz/2 bit per clock cycle} = 419.430 \text{ MHz}$$

Since DDR LVDS communication bus requires much less of expensive PCB space, the second approach is often considered as more cost efficient. However, it assumes existing on-chip parallel to serial conversion mechanism (circuit). This mechanism is called SerDes and allows conversion of parallel data word to the series of bits and vice-versa. This is the reason why most FPGA vendors have started to include hardware cores for multigigabit SerDes in their recent FPGA devices (e.g., Xilinx 7 Series FPGA [17] or ALTERA Stratix FPGA [18]). It is necessary to mention that utilization of the aforementioned SerDes units will increase the cost of the associated boards due to (1) additional power and ground layers, (2) special radio frequency (RF) signal layer(s) layout on the multilayer PCB, (3) SI simulation prior to PCB manufacturing, and (4) postmanufacturing test of actual bandwidth using relatively expensive instruments. Thus, there are certain trade-offs associated with utilization of parallel versus serial methods in designing on-board high-bandwidth buses. Quite often, the parallel synchronous approach is used for relatively short (up to 100 mm) buses using single-ended LVTTL lines. For longer on-board lines and frequencies exceeding 200 MHz, the LVDS-based custom buses are more cost efficient. For interboard communication, utilization of multigigabit SerDes cores became a "de facto" standard solution in recent years. SerDes devices can be built-in to the FPGAs or can be separate stand-alone devices for interboard communication. The optical (light emitting diode [LED] or laser-based) or RF coaxial cables are the usual communication media for SerDes devices. In both cases, the bandwidth can exceed 10 Gb/s. However, utilization of optical cables is preferable in many cases specifically when electromagnetic interference could be an issue.

4.11 Summary

This chapter was devoted to the problem of creation of configurable and reconfigurable communication links between the on-chip PEs and organization of interface between the on-chip level of the RCS and its on-board level. The focus was put on practically utilized approaches in the organization of the on-chip communication infrastructure in modern FPGAs. Also, principals and organization of IOEs were discussed focusing on (1) organization of DCI, (2) tri-stating of OBUFs for bus interfacing, and (3) bandwidth increase using DDR organization of data transmitters and receivers. Another important aspect discussed in this chapter was the reduction of electromagnetic noise in the high-speed data-transfer lines. Single-ended and differential signal lines organization have been considered including associated electrical standards for the aforementioned communication lines (e.g., LVTTL, LVDS). Nowadays, these standards for signaling are included in most of the recent FPGA devices and can be configured for proper organization of on-board communication channels between FPGA(s) and other devices deployed on the PCB. This is important for organization of high-bandwidth interface to on-board memory devices that usually utilize differential signaling combined with DDR data-transfer organization. The concept of DDR was presented in detail due to its wide use in many high-bandwidth peripherals like DDR-SDRAM memory. The configurable coarse-grained interface elements like SerDes were also briefly discussed.

Exercises and Problems

Exercises

1. List the reasons that make the communication infrastructure more flexible at the on-chip level in comparison to the on-board or multi-board levels of the RCS.
2. What are the on-chip components in modern PLDs to be interconnected by the on-chip communication link and/or buses?
3. Define the components that must always be part of a programmable routing element.
4. How many combinations of interconnect between four communication lines can provide the six-pass-transistor-based PSE depicted in Figure 4.1? Which switches out of S1, S2,..., S6 should be closed to provide connection between lines *LW*, *LS*, and *LE* isolating *LN*? How many bits of configuration memory are necessary to have for this PSE?

5. Why in recent FPGA architectures are the different lengths of communication lines utilized? Cite examples.
6. Why must IOEs as signal adaptors be included in PLD architecture?
7. Define the four main circuits and their functions to be part of the IOE in PLD.
8. Why are power sources for the on-chip logic circuits and system levels (interface to the PCB circuits) of the same PLD different?
9. Define reasons why multiple data-transfer standards and associated line termination techniques should be utilized in PLDs for each data transaction line.
10. In what case does the dynamic organization of inter-PLD communication more efficient than static point-to-point links between PLDs?
11. Can multisourcing cause permanent hardware failure in OBUF of the PLD? If yes, what is the reason?
12. What is the most efficient method to prevent multisourcing in shared data-transfer lines? Define the conditions for the utilization of this method.
13. Determine the main components of generic IOB with controlled impedance and the ability to avoid multisourcing in shared data lines.
14. What is the difference between single-ended and differential signaling?
15. What are the pros and cons for communication links/buses based on TTL standards?
16. What are the pros and cons for communication links/buses based on LVDS standards?
17. How many bits per one clock cycle are possible to transmit over one communication line (one wire in case of LVTTL and two wires in case of LVDS)?

Problem

If the required peak bandwidth of the data bus connecting FPGA with memory block is equal to 2 GB/s, calculate the number of data lines in this bus in case of

1. Single-ended LVTTL synchronous bus with clock frequency equal to 67.2 MHz
2. Differential LVDS synchronous DDR bus with clock frequency equal to 536.9 MHz

References

1. K. Duong, S. M. Trimberger, and A. Mehrotra, Programmable single buffered six pass transistor configuration, Patent: US 5600264 A, Publication: February 4, 1997.

2. B. K. Britton, D. D. Hill, and W. A. Oswald, FPGA with distributed switch matrix, Patent: US 5396126 A, Publication: March 7, 1995.
3. Xilinx Inc., *Spartan-6 FPGA Configurable Logic Block*, User guide, UG384 (v1.1), February 23, 2010.
4. S. A. Hronik, Double data rate synchronous SRAM with 100% bus utilization, Patent: US 20030167374 A1, Publication: September 4, 2003.
5. W. Hill and P. Horowitz, *The Art of Electronic*, Cambridge University Press, New York, 1989, pp. 495–497.
6. W. H. Hayt Jr., *Engineering Electromagnetics*, 5th ed., McGraw-Hill, New York, 1989.
7. D. Brooks, *Signal Integrity Issues and Printed Circuit Board Design*, Prentice Hall, Upper Saddle River, NJ, 2003.
8. Xilinx Inc., *Spartan-6 FPGA Select IO Resources*, User guide, UG381 (v1.6) February 14, 2014.
9. ALTERA Corp., *MAX 10 General Purpose I/O*, User guide, UG-M10GPIO, 2015.06.10.
10. Xilinx Inc., Using Digitally Controlled Impedance: Signal Integrity vs. Power Dissipation Considerations, Application Note: Virtex-5, Virtex-4, Spartan-3 Devices, XAPP863 (v1.0), June 1, 2007.
11. S. M. Nolan and J. M. Soltero, *Understanding and Interpreting Standard-Logic Data Sheets*, Texas Instruments, Dallas, TX, Application Report SZZA036B, May 2003.
12. Texas Instruments, SSTL for DIMM Applications, SCBA014, December 1997.
13. National Semiconductor, *LVDS Owner's Manual Low-Voltage Differential Signaling*, 3rd ed., Texas Instruments, Dallas, Texas, 2004.
14. Xilinx Inc., *ZC706 PCIe Targeted Reference Design (ISE Design Suite 14.3)*, User guide, UG963 (v1.1), July 3, 2013.
15. K. Palanisamy and R. Chiu, High-Performance DDR2 SDRAM Interface in Virtex-5 Devices, Xilinx Application Notes XAPP858 (v2.2), September 14, 2010.
16. ALTERA Corp., Interfacing DDR bSDRAM with Stratix II Devices, Application Note AN-327-3.2, September 2008.
17. Xilinx Inc., *7 Series FPGAs Select IO Resources*, User guide, UG471 (v1.5), Chapter 3, May 15, 2015.
18. ALTERA Corp., *Altera LVDS SERDES IP Core*, User guide, UG_Altera_lvds, August 18, 2014.

5

System-Level Organization of the FCR

5.1 Introduction

In Chapters 3 and 4, the organization of reconfigurable processing elements (PEs) and reconfigurable communication infrastructure for PEs at the on-chip level of the field of configurable resource (FCR) was discussed. However, FCR can be extended to the system level in the reconfigurable computing system (RCS). The system level of the FCR assumes onboard or multiboard hardware implementation in the form of multiple programmable logic devices (PLDs) accommodating reconfigurable processing units (RPUs). Each reconfigurable processing unit (RPU) being implemented in the form of system-on-programmable chip is optimized for processing the specific task segment(s). In most of the practical examples of reconfigurable systems, RPUs are implemented in the field programmable gate array (FPGA) devices or complex programmable logic devices (CPLDs). However, some application-specific RPUs can be implemented on the hardware basis of coarse-grained reconfigurable arrays (CGRAs), as was discussed in Chapter 3. The aspects associated with the configuration of RPUs at the on-chip level of RCS will be discussed in the following chapters.

In the hybrid reconfigurable systems, the programmable processing elements (PPEs) can also be included in the FCR at the on-chip and system levels. In the case of the system level, the PPUs in most of the cases are macrofunction-specific programmable processors (e.g., DSPs, compression–decompression [CoDec] processors). These RPUs and PPUs should be interconnected for information exchange at the system level. Thus, the set of RPUs, PPUs, and communication infrastructure between them at the onboard or multiboard level represent the *FCR on the system level of the RCS*.

There are several reasons motivating the extension of the FCR from single to multiple PLDs and PPUs allocated on the same printed circuit board (PCB) or multiple PCBs. The main reasons for the expansion of FCR to more than one PLD are as follows:

- The application cannot be spatially mapped in a single PLD (e.g., FPGA device) due to limited resources in the available PLDs.

- It is necessary to conduct partial reconfiguration in the FCR during task execution when the available PLDs cannot provide this feature at the on-chip level. In other words, it may be necessary to reconfigure some RPUs when others cannot interrupt data-processing. Thus, the reconfiguration process should be distributed between several PLDs.
- The reliability constraints require parallel execution of the same segments of the task on different RPUs. The triple modular redundancy (TMR) solution can be considered as one of the examples for this case.

In the aforementioned cases, the inter-PLD communication infrastructure should be deployed at the system level of the FCR and can be considered as a *network between PLDs* as network nodes. This organization also depends on devices providing the reconfiguration support for the FCR.

Therefore, this chapter is focused on the system level of FCR organization specifically on the organization of PCB level of communication infrastructure. Pros and cons of static point-to-point links will be compared with dynamically reconfigurable links between processing units.

5.2 FCR Organization Based on Static Links between PLD-Nodes

The RCS with multiple PLDs composing FCR on system level of architecture organization was usual for early RCS architectures due to the relatively small amount of configurable resources in the FPGA/CPLD devices in the 1990s and early 2000s [1]. However, further progress in deep-submicron complementary metal-oxide semiconductor (CMOS) technologies significantly increased the available configurable logic resources in the FPGA devices. Thus, the reason for multi-FPGA implementation of the FCR on system level is not as strong as it was in the early ages of RCS development. On the other hand, there are still many cases when complex multimodal applications require multiple large FPGA devices or CGRA devices to accommodate the required PRUs and satisfy the requested performance and reliability constraints. The aforementioned applications may require two or more PLDs to accommodate all hardware resources. The organization of PCB-level communication links directly depends on the number of PLDs and available interface elements in each PLD. In the case of two PLDs on a board composing PCB level of the FCR, the usual organization of the communication between these two FPGAs is a bidirectional link (e.g., reconfigurable application specific processor (RASP) "blade" module [2]). This link can be organized as a set of simplex, half-duplex, or full-duplex (bidirectional) communication channel(s) allowing communication between PLDs. The

FCR composed by two programmable logic devices

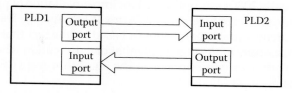

FIGURE 5.1
Duplex point-to-point printed circuit board–level communication link between two field programmable gate array devices.

solution depends on communication bandwidth requirements and could be organized in the form of serial or parallel data-transfer lines between these FPGA devices. The example of point-to-point (full-duplex) simple network is shown in Figure 5.1 representing physical link statically organized between two PLDs composing the FCR.

The bandwidth of this type of links should satisfy the highest data rate for data exchange between computing circuits allocated in PLD1 and PLD2 at any mode of operation. Otherwise, the link can become the bottleneck of the RCS architecture limiting the performance of the system. Since the peak bandwidth may not be reached all the time, there are certain hardware and timing overheads assumed in this organization of communication between FPGAs.

On the other hand, this is the simplest possible solution. The organization of point-to-point links between more than two PLDs will require additional input/output (I/O) ports. The number of those additional ports depends on the topology of interconnects between PLDs. In turn, this topology depends on the specifics of data-flow required for the application. For instance, pipeline organization of RPUs allocated in multiple PLDs requires an additional I/O port for each PLD, as shown in Figure 5.2. The pipeline could be unidirectional or bidirectional. The bidirectional inter-PLD links can provide feedback signals as well as recurrent data. The pipelined organization of data-path is an effective solution for conventional computers and RCS [3]. This organization is effective in the case when variation of data-flow rate is relatively small for all modes of operation. Otherwise, resource utilization factor for interface/communication resources can be reduced.

Pipeline topology of PLD interconnection in the FCR

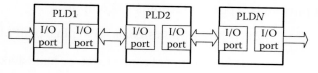

FIGURE 5.2
Pipeline-like network in multi–programmable logic device reconfigurable computing system architecture.

Furthermore, the micro-architecture of some families of FPGA devices in late 1990s and early 2000s was optimized for implementation of pipelined data-paths of SoCs (e.g. Xilinx Virtex-E and Virtex-II FPGAs [4]). The efficiency of the pipelined architecture organization is based on the fact that many classes of applications assume multistage sequential data-execution where each stage is associated with a certain segment of the algorithm (e.g., DSP, video- and packet-processing applications).

It is necessary to mention that PLDs deployed in different pipeline stages may or may not be equal to each other. Instead, they usually reflect the needs of certain parts of algorithm(s) and data structure. In addition, the bandwidth of each inter-PLD link may be different, reflecting specific data-transfer requirements between segments of an application deployed in different PLDs. In general case, the number of I/O ports assigned for inter-PLD communication is doubled in comparison with simple dual-PLD communication link. On the other hand, this number of I/O ports does not depend on the number of PLDs included in the pipeline. Nevertheless, there could be applications where pipelined processing is not an effective solution. This is usual for the tasks where several segments of the application should be executed in parallel and at the same time need to exchange data between each other. In general case, each PLD should have the direct link to all other PLDs. This solution was often used in RCS architectures [5]. The advantage of such organization of the onboard communication network is the highest flexibility and thus highest possible performance of the RCS. The major disadvantage of this static network organization is the complex multilayer PCB layout and a lot of interface resources arranged for data exchange between PLD-nodes. Practically, these resources being designed to accommodate the highest bandwidth may be underutilized. In general, if the aforementioned FCR consists of NPLDs, the number of I/O ports in each PLD will need to have $(N - 1)$ I/O ports to provide point-to-point communication links. For example, if the FCR consists of 4 PLDs, each PLD will require $4 - 1 = 3$ I/O ports providing direct communication between any pairs of PLDs. The organization of this type of network is depicted in Figure 5.3. Obviously, the number of I/O ports in each PLD dedicated for inter-PLD communication is three times more than that in dual-PLD architecture.

As an example, let us consider bidirectional full-duplex organization of 32-bit data-transfer links. In this case, the number of PLD pins dedicated for inter-PLD communication is as follows:

Number of pins = 3 I/O ports × (32 inputs + 32 outputs

+ 2 synchronization signals = $\underline{198 \text{ pins/per PLD}}$

+ pins for additional control and/or handshaking signals

However, the associated PCB layout should provide six separate inter-PLD links and thus

FCR composed by four programmable logic devices

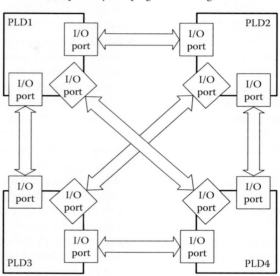

FIGURE 5.3
Duplex point-to-point onboard communication network for four programmable logic devices.

$$6 \text{ (inter-PLD links)} \times (64 \text{ data-transfer lines}$$
$$+ 2 \text{ synchronization lines}) = 396 \text{ lines}$$

Thus, doubling the number of PLDs at the onboard level has increased by six times the number of communication lines on the PCB.

The next doubling of resources in the FCR will cause eight fully interconnected (with each other) FPGAs. In turn, this FCR organization will require 462 pins for each PLD and 28 inter-FPGA communication links deployed on PCB. The complete set of $28 \times (32 \text{ bit input} + 32 \text{ bit output})$ I/O lines + associated synchronization lines will require 1848 single-ended lines. This will significantly increase the cost of PCB design, layout, and production.

Obviously, the communication network based on the set of point-to-point links provides the highest possible flexibility in the organization of information exchange between different PLD-nodes. However, network organization based on static point-to-point links between all PLD-nodes causes a dramatic increase in the number of I/O pins and associated interface elements in each PLD. In addition to that, significant growth of complexity of the PCB limits the utilization of this approach for the multi-PLD organization of the FCR on the other hand.

To overcome this problem, instead of the aforementioned general-purpose organization of PLD interconnects, more application-specific approach may simplify the communication network on the PCB level of the FCR. As an example, it is possible to consider the spatially parallel computing architecture where several PLDs are executing *independent segments of the task* simultaneously. This type of topology is similar to array processing systems. However, in contrast to conventional array (or vector) processors providing parallel execution of independent data-elements by the same segment of an algorithm, each PLD may accommodate its own RPU optimized for dedicated task segment. Therefore, this "array-like" organization of topology allows simultaneous execution of independent tasks, segments of tasks, or large blocks of data. Furthermore, each PLD could be configured/reconfigured dedicated to RPU independently and thus provide higher flexibility for the FCR. This topology is shown in Figure 5.4.

Usually, the array organization of multi-PLD FCR is used for acceleration of data-execution in computation-intensive segments of the application. As an example, it is possible to mention data encryption/decryption, CoDec, error correction, etc. Distribution of task segments on RPUs as well as coordination of data-execution processes running in parallel in the PLD-nodes are usually conducted by a separate subsystem. This subsystem could be based on PPU-centric sequential microprocessor or another PLD dedicated just for that task.

In case of very complex multimodal and multistream applications, the array topology can be combined with the pipeline topology resulting in 2D array topology of FPGA-node interconnects. This type of network shown in Figure 5.5 was widely used in systolic array architectures of parallel computers (e.g., ILLIAC IV [6]) as well as early RCS (e.g., virtual computer [7]).

In this type of topology, each stream of data can be assigned to a certain chain of PLD-nodes. This allows a more flexible distribution of segments of task(s) and parallel running in the pipelined processing circuits accommodating different data-streams. Another big advantage of this topology is the ability for partial reconfiguration of the FCR based on nonpartially reconfigurable PLDs. The close placement of PLD-nodes to each other allows simplification of the PCB layout because short single-ended lines usually can be allocated on the same signal layer. This organization was quite popular when the amount of logic resources in FPGA devices was limited and

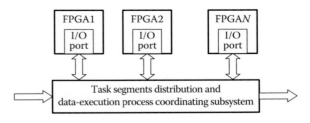

FIGURE 5.4
Array topology of field programmable gate array interconnections in the field of configurable resource.

2D topology of PLD interconnection in the FCR

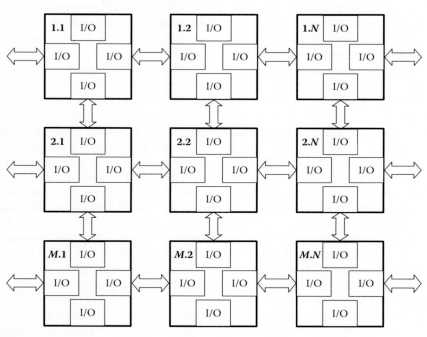

FIGURE 5.5
A 2D systolic array communication network for multi–programmable logic device field of configurable resource.

FPGA devices did not allow partial reconfiguration at the on-chip level. In recent years, this type of topology is not so popular due to the elimination of the aforementioned reasons. Nowadays, most FPGA vendors provide large FPGAs with the ability for partial reconfiguration in their tile-based architecture and on-chip 2D routing structure (e.g., Xilinx Virtex-5/6 FPGAs or Xilinx 7 family of FPGA devices). Thus, the implementation of 2D systolic array architecture is more effective at the on-chip level rather than at the system level due to the reduction of PCB layout complexity.

5.3 Organization of Dynamic System-Level Network in the Multi-PLD FCR

The aforementioned static topology of links between onboard PLD-nodes as well as other nonreconfigurable components is effective when all links are used most of the time while RCS is active. Different topologies of static communication networks have been deeply investigated for different parallel computer architectures [8]. However, if application does not utilize static links, resource utilization factor for system level of communication resources can be significantly reduced. In such cases, the dynamic links between PLDs can be considered as an effective solution.

Let us consider the 4-PLD-based FCR depicted in Figure 5.3. In case of full-duplex I/O ports (32-bit input + 32-bit output + 2 synchronization) and single-ended transmission lines, each PLD will need to use three I/O ports with $3 \times (32 + 32 + 2) = 198$ I/O pins and associated interface elements. Let us assume that during the mode M_i, each PLD-node receives source data-stream from the sensors or other PLD and generates the output data-stream to another PLD. In other words, in this mode the pipelined processing system is combined out of all the existing PLD-nodes. In turn, each PLD uses one input port with 32-bit data + 1 synchronization signal = 33 I/O elements and one output port with the same 33 I/O elements. Thus, only 66 I/O elements are used in this mode when other $198 - 66 = 132$ I/O elements are not in use. Therefore, only $66/198 \times 100\% = 33.3\%$ of communication resources are used in this mode. This example shows that it is quite difficult to keep a high level of resource utilization factor for the static communication network designed for all possible topologies of interconnects between PLD-nodes. At the same time, communication links between PLDs are relatively expensive hardware resource requiring area of PCB, number of I/O elements in each PLD, and associated design and production costs. This resource also requires energy for these I/O interfaces. Therefore, the increase of efficiency of inter-PLD network can be reached by dynamic configuration of links topology according to the data-flow structure of the task in current mode. This topology of links can be configured in special PLD, which serves as a local router for the set of PLDs deploying RPUs. This router can be built on the base of cross-point switches [9] or multiplexers. An example of dynamic communication network for four PLDs is shown in Figure 5.6.

For example, if Xilinx Spartan 3S (XC3S400-4FGG456) is used as an interconnect switch (router), it can provide 264 I/O pins or 4×64-bit I/O ports plus 2 extra pins for control/synchronization signals per port. The cost of this additional onboard communication FPGA-based switch is approximately \$40. It is necessary to mention that in most cases the reason for the expansion of the FCR from single to multiple FPGAs is the lack of logic on-chip resources in the single FPGA. Thus, each FPGA is assumed as the largest device in its family. However, large Xilinx FPGA devices are relatively expensive units. For example, the cost of Xilinx Virtex-6 FPGA XC6VLX240T-1FF784C with 400 I/O pins is close to or more than \$2000. Therefore, the overhead cost of additional \$10/port is negligible in comparison with the cost of the large FPGA of the same vendor.

Due to the specifics of production, the cost of I/O resources in large FPGAs is relatively higher than the cost of the same I/O resources in small FPGAs.

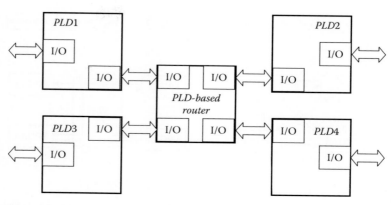

FIGURE 5.6
Communication network organization with central programmable logic device–based router.

Taking the earlier example, it is possible to see that the difference in device price between Xilinx XC6VLX240T-1FF784C with 400 I/O pins and the same device XC6VLX240T-1FF1156C with 600 I/O pins is close to $300 per device or approximately equal to $1.5 per I/O pin. The difference in price between XC6VLX240T-1FF1156C with 600 I/O pins and XC6VLX240T-1FF1759C with 720 I/O pins is close to $340 or equal to $2.8 per I/O pin. In contrast to that, the cost of I/O resources in the aforementioned Xilinx Spartan 3S (XC3S400-4FGG456) providing 264 I/Os for $40/FPGA device is equal to 15 cents per I/O element or $9.7 per 64-bit I/O port. Therefore, the implementation of dynamic communication switch in small FPGA can be a cost-effective solution in case of multiple onboard FPGA devices hosting RPUs. In this case, the cost and dimensions of these large FPGAs can be minimized. Taking the aforementioned example, the Xilinx XC6VLX240T FPGA is available in packages providing 400, 600, and 720 I/O pins. The difference in dimensions, cost of the device, and cost of the area of the supporting board directly depends on the aforementioned packages. For the aforementioned example, the largest package FF1759 (720 I/Os with dimensions = 42.5 mm × 42.5 mm) requires 2.15 times larger area than the smallest package for the same device FF784 (400 I/Os with dimensions = 29 mm × 29 mm). Therefore, the *cost-efficiency of multi-FPGA FCR designs can be increased significantly if the number of I/Os in the large FPGA devices is minimized.*

Now, let us analyze the efficiency in the utilization of communication resources for the FCR with four FPGA devices interconnected via central programmable switch matrix (PSM) deployed in small FPGA. The comparison with FCR organization based on statically interconnected four FPGAs (as shown in Figure 5.3) is illustrated in Figure 5.7.

Let us consider the example of application consisting of four consecutive task segments: TS_1, TS_2, TS_3, and TS_4. The data-flow structure of this application is shown in Figure 5.7a. Each RPU executing an associated task

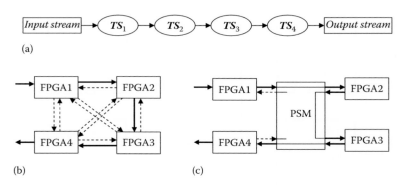

FIGURE 5.7
(a) Data-flow structure of the application, (b) topology of active links (bold lines) between RPUs in case of fully interconnected FPGA-nodes, and (c) topology of links configured in programmable switch matrix (PSM) deployed on dedicated FPGA device.

segment can be mapped to any FPGA device composing the FCRs of the RCS. If task segments are allocated in the FPGA nodes according to their numbers, then the topology of links should reflect data dependencies between the aforementioned segments. Figure 5.7b shows the implementation of the aforementioned application on the FCR based on fully interconnected full-duplex communication network. The bold lines are representing the active links transferring the data-streams from one FPGA node to another composing the pipeline of FPGA-based processing units. The dotted lines in Figure 5.7b show the existing (in infrastructure) but nonutilized communication links. Thus, out of 12 32-bit I/O ports and associated point-to-point links, only 3 inter-FPGA ports and associated links are actively utilized by application. Thus, the utilization of communication resources is equal to $3/12 \times 100\% = 25\%$.

The alternative solution is the FCR with the network based on switching matrix deployed in the small FPGA. The same pipeline of four processing units is composed by programming of interconnects in the FPGA-based PSM, as depicted in Figure 5.7c. Each processing unit allocated in the FPGA requires only one I/O port for inter-FPGA communication instead of three that are used in fully interconnected network shown in Figure 5.7b. It is also possible to see that out of eight I/O ports and associated onboard links, only six are in use. In this case, the resource utilization of the onboard communication resources is equal to $6/8 \times 100\% = 75\%$, which is 3 times higher than in the case of statically configured network. Obviously, FPGA-based PSM allows the implementation of almost any topology of links at the onboard level of the FCR. Thus, this approach significantly increases flexibility of the onboard network in comparison with static topology of point-to-point links discussed earlier in Section 5.2.

In addition to that, it is necessary to mention that the width of input and output ports can be easily customized in the case of central PSM unit. Indeed, since I/O elements reserved for each I/O port are identical and programmable, the number of I/O elements combined and programmed as input port or output port could be changed according to the application needs. For example, if 64 I/O elements are reserved for the I/O port in the FPGA, it is possible to program 32 I/O elements as inputs and other 32 as outputs for a certain mode of operation. In other modes of operation, the same 64 I/O elements can be divided in 16 outputs and 48 inputs. In the case of static point-to-point links, the aforementioned customization is also possible but only between two FPGA devices interconnected by the physical onboard link. In the case of PSM unit, the output port of the FPGAi can be connected to the input of the FPGAj, when the input port of the FPGAi receives the data-stream from the output port of the FPGAk. Thus, another aspect of interconnectivity gives more flexibility in the network based on dynamically reconfigurable links.

The aforementioned schemes of FPGA-node interconnection assumed that the processing data rate in the FPGAi is equal to the bandwidth of I/O ports of the associated FPGA. If the bandwidth of the output port is higher than the performance of data-processing circuits in this FPGA, the output port will be underutilized. In an opposite case, when the processing data rate is higher than the available bandwidth, the buffering circuits should be included to the design (e.g., FIFO or LIFO buffers). In the case of central PSM unit, these functionalities could be allocated in communication management circuits. These circuits may perform buffering of inter-FPGA information. In addition, balancing of data rates and I/O bandwidth for different FPGAs can also be performed in this FPGA-based central communication unit (CCU). Thus, these mechanisms may significantly increase the utilization of communication resources for multimodal and/or multitask applications.

Let us consider now the example of array-like multi-FPGA topology of communication links shown in Figure 5.4. The data-flow structure of this application is presented in Figure 5.8a. For simplicity, let us assume that the application consists of four segments, TS_1, TS_2, TS_3, and TS_4. The main part of the algorithm is included in segment TS_1 and contains mostly algorithmic-intensive parts of the application. Computation-intensive parts of the application are allocated in segments TS_2, TS_3, and TS_4. These segments are getting requests and data from the main task segment TS_1. After completion of data-execution according to TS_i, $i = 2, 3, 4$ returning results to TS_1.

This kind of application is common for DSP and video-processing applications, where certain parts of the task are executed on so-called hardware accelerators. The configuration of communication links, associated buffers, and multiplexers is shown in Figure 5.8b.

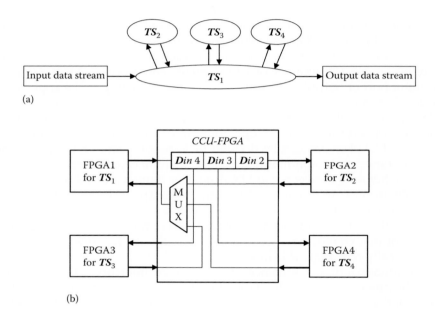

FIGURE 5.8
Implementation of the application in the FCR with configurable communication unit. (a) Data-flow structure of application and (b) configuration of communication links, buffers, and multiplexors in CCU.

This figure illustrates the concept of inter-FPGA communication via central configurable communication unit (CCU). The CCU may include

1. Memory units (block RAM or FIFO/LIFO buffers) dedicated to RPUs allocated in FPGAs and interfaced with this memory unit(s)
2. Multiplexers/demultiplexors allowing switching of input port of one FPGA (e.g., FPGA1 in this example) to outputs of other FPGAs sequentially
3. Set of links between the CCU components and I/O ports of all FPGA nodes in the FCR

All these resources require control and process synchronization circuits, and thus, the CCU deployed on PCB level must perform these functions as well. Furthermore, the control and synchronization functions usually are distributed between processing units deployed in the FPGA nodes and interconnecting CCU. In addition to that, CCU can provide communication with peripheral devices as well as interaction with other CCUs. This allows aggregating several CCUs in more complex network. In addition, CCU can be used as the majority gate or voter if the TMR type of computing is required for fault-tolerant data-execution in dependable segment(s) of the application.

If the voter can be affected by radiation effects or other reasons, the voter circuits also can be triplicated in the CCU.

In sum, the additional FPGA-based CCU can perform multiple functions from simple data-transactions/distribution between processing nodes to complex buffering/multiplexing the data-streams as well as TMR and error correction functions.

5.4 Organization of the Hybrid FCR with Programmable Processing Elements

There are several reasons when the external (onboard) PPEs can be a more effective solution than the embedded PPE. The most important and usual reasons are the following:

1. Separation of the low-bandwidth interface devices from the high-performance part of the system
2. Simplification of design using off-the-shelf (OTS) function-specific PPEs (e.g., floating-point DSP, CoDec processors)
3. Increasing the reliability by distribution of the same functionality on different off-chip PPEs (e.g., TMR)

Let us consider each of these cases in detail in the following sections.

5.4.1 Separation of the Low-Bandwidth Interface Part of the System

In many cases, the cost of the system can be reduced by using FPGA devices without embedded PPEs in cases when PPEs are utilized for interfacing to the low-speed peripheral devices. Instead, the low-cost microcontrollers (e.g., 8/16-bit Microchip PIC [10] or Avnet AVR [11]) can be used more cost-efficiently for the operations that do not require high-performance data-acquisition, data-conversion (e.g., low-rate mixed signal operations: Analog-to-digital conversion (ADC) or digital-to-analog conversion [DAC]), preprocessing, and/or interfacing to the low-bandwidth peripherals (e.g., keyboard or keypads, PWM motor control). These controllers can be attached to the host FPGA via serial (e.g., SPI or I2C) or 8/16-bit parallel bidirectional ports directly. This way of interfacing low-cost microcontrollers to the FPGA-based FCR is shown in Figure 5.9 and does not require many I/O elements in the target FPGA. However, this solution is effective in cases of very few peripheral controllers. Otherwise, this solution may require too many I/O pins and associated I/O buffers in the FPGA. On the other hand, relatively low data rate of 8/16-bit microcontroller's ports will not be able to utilize effectively the

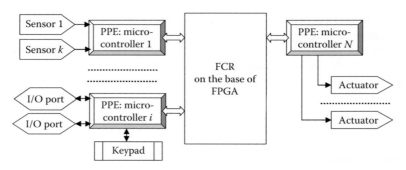

FIGURE 5.9
Direct interfacing of external microcontrollers to the field programmable gate array–based field of configurable resource.

high-bandwidth and relatively expensive low voltage transistor-transistor logic (LVTTL) or low voltage different signaling (LVDS) I/O blocks (I/OB) of the FPGA. This aspect of interfacing was discussed in detail in Chapter 4.

Alternatively, the PPEs can be interfaced to the FPGA via low-cost CPLD/FPGA-based communication unit performing the role of peripheral controller, as shown in Figure 5.10. This solution minimizes the number of FPGA pins,

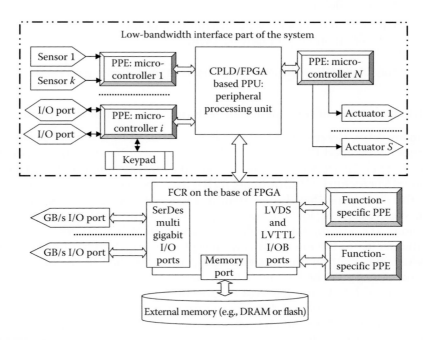

FIGURE 5.10
Interfacing of external microcontrollers, function-specific processors, and external memory devices to the field programmable gate array–based field of configurable resource.

I/O resources, PCB layers, and thus system cost. Therefore, more FPGA pins can be utilized more effectively for communication with peripherals, which require high-bandwidth communication (e.g., double data rate synchronous dynamic random access memory, flash memory units, and/or function-specific processing units). This approach also allows reduction of the system cost:

1. Reduction of the component cost. In most cases, peripheral interface controllers may cost units of dollars when the equivalent cost of on-chip PPEs is, in order of magnitude, more expensive.
2. Cost reduction of the development platforms. The cost of the development platforms for 8/16-bit microcontrollers can range from tens to hundreds of dollars in comparison with thousands of dollar cost of the evaluation/development platforms for the FPGAs with embedded PPEs.
3. Reduction of the design time and therefore design cost. The design of PPE and associated software or firmware based on 8-bit or 16-bit external OTS microcontrollers is less than that in the case of utilizing on-chip hardcore or soft-core PPEs.

5.4.2 Simplification of Design Using Macrofunction-Specific PPEs

Quite often, the application includes *macrofunctions* that are in constant use in most or even all modes of operation. As examples, it is possible to mention complex DSP macrooperations, functions requiring floating-point operations, CoDec operations (e.g., MPEG or JPEG), etc. Implementation of these macrofunctions can be done by either of the following:

1. Software solution in a form of loadable to the PPE program
2. Hardware implementation in a form of application specific integrated circuit (ASIC) optimized for macrofunction or intellectual property (IP) core (configuration bit-file) optimized for a certain type of the FPGA
3. Hybrid circuit consisting of function-specific PPE plus hardware accelerator

Software implementation can be used for the embedded PPE or external processor. However, hardware or hybrid implementation always assumes external macrofunction-specific ASIC.

The use of software components may be the cheapest and the simplest way for macrofunction implementation. However, the performance requirements may not be satisfied due to the sequential nature of data-execution in conventional PPEs. As an alternative, the external ASIC(s) can be considered as competitors for IP cores. The analysis of the cost-efficiency should be conducted according to specification constraints, including system cost, design time, and cost of preproduction preparation. In case when the external macrofunction-specific processors are selected as the cost-effective solution, interfacing to the FPGA-based FCR in most cases is done directly. These interfaces usually utilize LVDS or LVTTL I/OB embedded into the FPGA architecture or serializer/deserializer on-chip devices, as shown in Figure 5.10. The details of this type of interfaces were discussed in Chapter 4.

The utilization of external macrofunction-specific processors provides the following benefits:

1. Reduction of design time and cost and time to market because a big portion of the design of the required macrofunction-specific circuit is embedded in OTS ASICs.

2. Minimization of component cost. The macrofunction-specific processor in a form of ASIC may cost from units to tens of dollars per function in comparison with thousands or even tens of thousand dollars for respective IP cores for the FPGA devices. Additionally, IP core may require a large portion of the FPGA resources, which by itself may cost more than the competitive ASIC performing the same macrofunction.

3. Simplification of integration and debugging. The external macrofunction-specific processors are usually mass-produced devices. Therefore, their interfaces are done according to industry standards. In addition to the above, self-testing and debugging circuits usually are embedded to these processors. This simplifies system integration process.

The practical implementation of the aforementioned principles of organization, the hybrid FCR can be considered in the architecture of the JPEG2000 CoDec and transport stream composition platform, 4Vision-JPEG2000 for stereo-panoramic video-applications. This platform is depicted in Figure 5.11.

This multimodal stereo panoramic video-processing system contains statically reconfigurable FCR that is deployed in two FPGA devices (Xilinx Spartan) and allows the following:

1. Parallel receiving of four XGA/720p video-streams from the video-acquisition subsystem

2. Parallel dynamic distribution of video-streams to JPEG-2000 video compressors or decompressors according to the requested mode of operation

3. Combining the compressed video-streams into the asynchronous serial interface (ASI) transport stream or separation of the ASI transport stream to JPEG-2000 compressed video-streams according to the mode of operation

4. Automated generation of the required support information for ASI packet headers, etc.

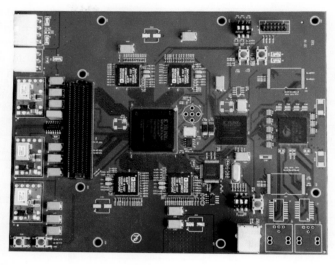

FIGURE 5.11
4Vision-JPEG-2000 video CoDec platform. (Courtesy of Embedded Reconfigurable Systems Lab (ERSL), Toronto, Canada.)

The block diagram of the architecture of this system is presented in Figure 5.12.

The central FPGA is directly linked to four JPEG-2000 CoDec ASICs (ADV212 by Analog Devices Inc.), each of which performs CoDec of individual video-stream. At the same time, this FPGA is connected by a bidirectional link to the peripheral processing unit (PPU) deployed on the second FPGA. This PPU can continue nonstop operation for the periods when the central FPGA is in the process of mode switching. Thus, transport video-stream can be transmitted or received without interrupts. Therefore, this architecture allows individual reconfiguration of two parts of the FCR deployed in different low-cost FPGA devices.

If all the aforementioned functions are implemented in system-on-chip (SoC) containing multiple IP cores performing JPEG-2000 CoDec functions and noninterruptible operation with high-bandwidth peripherals, the large partially reconfigurable FPGA would be necessary to use in this design (e.g., Xilinx Virtex-5/6 FPGA). In this case, the FPGA device may cost a few thousand dollars. The solution based on dedicated JPEC-2000 CoDec devices allowed using the low-cost FPGAs and therefore reduction of component cost in order of magnitude as well as design, prototyping, verification, and debugging periods of time. In addition to that, the high-bandwidth interface is also implemented in standard-specific ASIC (e.g., HOTLink transceiver).

The reconfiguration of FPGAs in this architecture is conducted by configuration microcontroller according to the mode of operation. The microcontroller

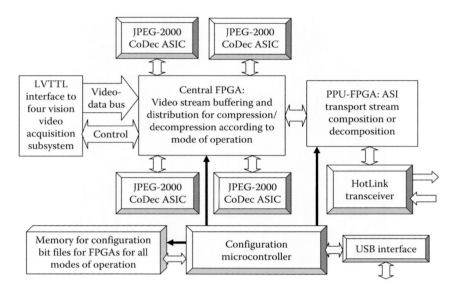

FIGURE 5.12
Architecture block diagram of 4Vision-JPEG-2000 video CoDec system.

receives commands for mode switching via USB interface and then loads the configuration bit-file from the flash memory to the target FPGA device. The content of library of configuration bit-files can be updated at any time by the same configuration controller over the USB interface.

Nonetheless, it is necessary to mention certain drawbacks for the utilization of macrofunction-specific processors in conjunction with FPGA-based FCR:

1. Lack of flexibility in design. In case when one of macrofunctions should be excluded or replaced by other macrofunction, it is very costly or practically impossible to do. Thus, the design solution based on the external processing ASIC chip is good for static macrofunctions, which must be in active mode all the time according to industrial standard.

2. Dependency on vendor's production. If the ASIC is discontinued in production or replaced by newer version, the redesign of the system which utilizes this ASIC may be needed. The problem can be mitigated if the replacement is pin-to-pin compatible with the previous device and the required macrofunction is in set without any change in specification constrains. In usual practice, vendors try to keep informed all customers regarding discontinuation of their products quite ahead of time. In addition, they try to keep the pin-to-pin compatibility. However, the problem in general exists anyway.

3. Area overhead. As any additional component deployed on the board surface, the external macrofunction processor requires additional space on the PCB and routing layout, which may increase system dimensions, weight, package cost, etc.

5.4.3 Increasing the Reliability by the Distribution of Functionality on the Off-Chip PPEs

There are cases when some *mission critical functions* can be affected by environmental effects or production and exploitation specifics (e.g., single-event effects caused by radiation, die aging, and manufacturing defects). In such cases, it may be an effective design solution to exclude these mission critical functions from the FPGA and allocate on more robust components (e.g., radiation-hardened chips, electromagnetic Interference (EMI)-shielded and thermostabilized containers). Very often, the TMR approach is used when the external processors performing these critical functions are combined in a group of three devices linked to radiation-hardened or radiation-shielded voting circuit. However, recently, the utilization of different on-chip self-restoration techniques for mitigation of transient or even permanent hardware faults in the FPGA devices became more popular and efficient, which can compromise onboard solutions.

5.5 Organization of the Multiboard Communication Network

As mentioned in the previous chapters, the entire RCS architecture and particularly the FCR should be considered in three levels: (1) on-chip level, (2) onboard (PCB) level, and (3) multiboard level. The last two levels, onboard and multiboard, can be considered as system levels of RCS.

The multiboard organization of the FCR and thus RCS may be needed in cases when:

- The aggregated amount of configurable resources allocated on single PCB is not enough to satisfy performance and/or reliability requirements of the workload
- The hybrid RCS architecture consists of different types of PEs or hardware accelerators optimized on different algorithms and/or data structures
- Statically configurable RCS assumes different plug-in boards for different classes of applications using the same processing platform
- Specifics of application dictate certain topology of PEs allocation

As an example of these multiboard RCS, it is possible to mention the following:

- Motherboard-based systems where central processing core combined with basic memory subsystem and general-purpose interface devices is combined with one or more RCSs utilized as hardware accelerators optimized for stream-processing tasks
- Multifunctional RCS with separate boards, each of which is designed to perform application functions associated with specific interfaces to complex peripheral devices, e.g., multivideo-stream acquisition and preprocessing

In most practical cases, the integration of multiple boards and associated organization of communication network assumes utilization of shared communication media-like buses or networks. There are many different ways for the aforementioned integration using different bus standards and network topologies and protocols. The organization of these buses and networks is similar to their organization in most of conventional computing systems and therefore is out of the scope of our consideration in light of the specifics of architecture organization of the RCS.

5.6 Summary

In this chapter, the organization of the FCR on the system level was discussed. Different types of communication network organizations have been observed for the FCR consisted by multiple RPUs, each of which is implemented as SoC at the on-chip level of the FCR. The FCR is deployed on PLDs (CPLDs, FPGAs, or CGRA devices).

It was shown that static inter-RPU links are cost-effective in cases when the inter-RPU data-flow variation is minimal. In other words, the condition for the cost-effective utilization of statically configured inter-RPU communication network is steady-state data-flow between PRUs. Otherwise, resource utilization may be significantly reduced creating relatively high communication overhead due to nonutilized PCB area and extra power in unused I/O elements.

Alternatively, dynamically reconfigurable communication network-based PSM or more complex CCUs could be an effective solution for the aforementioned multi-PLD FCR. The CCU being deployed in the small and relatively low-cost FPGA device can incorporate (1) routing resources, (2) buffers and local memory modules, and (3) multiplexers and demultiplexers providing highest flexibility in dynamic interconnection of different PLD ports with each other. Thus, the FCR equipped with CCU-based network can be effectively used for applications with big variations of data-flow between processing units.

In addition to that, the utilization of the CCU as the central interface device for multiple PPEs was considered for the hybrid organization of the FCR on the system level. The effectiveness of external PPUs included in the FCR on system level was also discussed with emphasis on their benefits and drawbacks.

Exercises and Problems

Exercises

1. Is it possible to implement the FCR only at the on-chip level? Prove your answer.
2. List the main reasons for the motivating extension of the FCR from single to multiple PLDs.
3. What are the pros and cons of the implementation of the FCR in single PLDs (e.g., FPGAs) compared to the implementation in multiple PLDs?
4. How many I/O ports should be reserved in each PLD for full-duplex data exchange with other PLDs in the case of static communication network in the form of
   a. Dual-PLD organization of FCR
   b. Pipeline organization of multi-PLD FCR
   c. N-PLDs fully interconnected with each other
   d. 2D systolic array topology
5. What is the main reason for the utilization of the dynamic communication network instead of static topology of links in multi-PLD RCS? When is this solution more cost-effective?
6. Why does the topology of interconnects between PLDs need to be changed? Can these changes be dynamic (during the task execution)?
7. What is the main reason for using the central PLD-based communication switch (router)? Why does this solution can reduce the cost of the RCS?
8. What is the main difference between PSM and CCU deployed in PLD dedicated for communication between RFUs? What are the main components in PSM and CCU?
9. What are the major benefits of the inclusion of CCU to the communication network in comparison with PSM, and in which cases this solution can be effective?

10. Describe the main reasons why the PPEs deployed at the system level of RCS can be a more effective solution than the embedded PPE.
11. In which cases does the separation of the low-bandwidth interface devices from the high-performance part of the system can be cost-efficient when the onboard PPEs (e.g., embedded microcontrollers) are used?
12. Define the benefits and drawbacks of utilization of external macrofunction-specific processors (ASICs) at the system level of the RCS architecture.

Problem

According to the example presented in Figure 5.7, analyze the efficiency of communication network utilization for the task segmented on four segments, TS_1, TS_2, TS_3, and TS_4. The *data-flow* structure of this task is shown in Figure P5.1.

1. Show the topology of active links between PLDs accommodating the associated task segments in case of (a) a fully interconnected multi-FPGA FCR and (b) an FCR with PSM based on FPGA.
2. Calculate the percentage of utilized communication resources as the ratio between active communication links and all links deployed on the system level of FCR for (a) a fully interconnected multi-FPGA FCR and (b) an FCR with PSM based on FPGA.

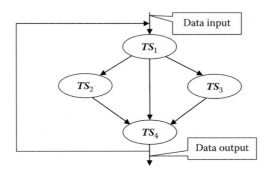

FIGURE P5.1
Data-flow graph of the task.

References

1. K. Compton and S. Hauck, Reconfigurable computing: A survey of systems and software, *ACM Computing Surveys*, 34(2), 171–210, June 2002.
2. Silicon Graphics Inc., Extraordinary acceleration of workflows with reconfigurable application-specific computing from SGI, White Paper, Silicon Graphics, Mountain View, CA, 2004.
3. L. Shannon, Reconfigurable computing architectures, in S. Hauck and A. DeHon (Eds.), *Reconfigurable Computing: The Theory and Practice of FPGA-Based Computation*, Morgan Kaufmann Publishers, Amsterdam, the Netherlands, 2008, Chapter 2, pp. 29–46.
4. Xilinx, *Virtex-II Platform FPGA Handbook, Part 1*, Xilinx UG002 (v1.0), December 6, 2000.
5. S. A. Guccione, Reconfigurable computing systems, in S. Hauck and A. DeHon (Eds.), *Reconfigurable Computing: The Theory and Practice of FPGA-based Computation*, Morgan Kaufmann Publishers, Amsterdam, the Netherlands, 2008, Chapter 3, pp. 47–64.
6. G. H. Barnes, R. M. Brown, M. Kato, D. J. Kuck, D. L. Slotnick, and R. A. Stokes, The ILLIAC IV computer, *IEEE Transactions on Computers*, C-17, 746–757, August 1968.
7. S. Casselman, Virtual computing and the virtual computer, in *Proceedings of the IEEE Workshop for Custom Computing Machines*, Napa, CA, April 5–7, 1993, pp. 43–48.
8. J. L. Hennessy and D. A. Patterson, *Computer Architecture: A Quantitative Approach*, 3rd ed., Morgan Kaufmann Publishers, San Francisco, CA, 2003, Chapter 8, pp. 788–877.
9. Xilinx, *Building Crosspoint Switches with CoolRunner-II CPLDs*, XAPP380 (v1.0), June 5, 2002.
10. T. Wilmshurst, *Designing Embedded Systems with PIC Microcontrollers: Principles and Applications*, 1st ed., Elsevier, Amsterdam, the Netherlands, 2010.
11. D. J. Pack and S. F. Barrett, *Atmel AVR Microcontroller Primer: Programming and Interfacing*, 2nd ed., Morgan & Claypool Publishers, San Rafael, CA, 2012, 246pp.

6

Configuration Memory and Architecture Virtualization in RCS

6.1 Introduction

In Chapters 3 through 5, the organization of configurable resources in on-chip and system levels was discussed. These resources include configurable processing elements (PEs), on-chip and system-level communication elements, and configurable interfaces contained in field of configurable resources (FCR). All the FCRs have mechanisms for programming their functionality by setting configuration bits to (1) configuration memory cells of respective programmable logic devices (PLDs), (2) configuration registers of coarse-grained resources, (3) look-up tables (LUTs) of fine-grained resources, and (4) memory blocks. These registers, LUTs, and memory blocks, associated with affiliated resources, have addresses in PLDs. Therefore, the complete set of the aforementioned memory elements dedicated for configuration or functionality programming can be considered as *configuration resources* of the reconfigurable logic device (e.g., FPGA, CPLD, CGRA). The content of these resources in any given moment of time determines the current functionality of components, information links between components, and procedures of their interactions. In other words, it determines the current architecture of the RCS deployed in its FCR. However, the concept of RCS assumes that the current architecture is not permanent. This architecture may change from time to time according to the changes of the mode of operation, changes in workload constraints, environmental conditions, etc. In turn, RCS should have additional memory units to store information associated with other possible RCS architectures or architecture variations. Furthermore, current set of possible architectures may need to be updated in the future. Therefore, the organization of the configuration resources in RCS can be considered as configuration memory subsystem separated from data memory subsystem. This subsystem should have certain memory hierarchy similar to the memory hierarchy in conventional computing systems with programmable procedure (e.g., RISC processors with virtual memory organization). As a result, in addition to data memory virtualization, hardware circuits associated with data-processing and data-transfer (i.e., PEs and communication links) also *can be virtualized*. The custom (dedicated) hardware components or even complete processing architectures can be represented by configuration bit-files and stored in different levels of the configuration memory hierarchy according to the probability of their activation. And only the components currently requested for operation are deployed in the FCR, thus becoming the actual hardware circuits. This allows significant minimization of hardware resources being utilized at a time in FCR. Therefore, virtualization of processing and communication resources allows significant minimization of cost, power consumption, and dimensions of the entire computing system.

The following sections will describe the general organization of configuration memory hierarchy and the concept of hardware circuit virtualization illustrated by quantitative examples of RCS.

6.2 Generic Organization of Configuration Memory Hierarchy in the RCS

As mentioned in Chapter 1, the computing process consists of two flows of information: (1) flow of data and (2) flow of control information. In accordance to this, memory organization must reflect this aspect on the computing process. In other words, the memory of the computing system should be divided (explicitly or implicitly) into two parts:

1. The part for storing the data-elements of the task
2. The part containing the control information of this task

In instruction-based computers, the control information in a form of procedural program is stored in the program memory, when the data are stored in data memory. In CISC type of computing architecture, memory stores both types of information (instructions and data-files). In these computing systems, only the program counter in the central processing unit (CPU) distinguishes the type of information in the addressed memory cell. The acceleration of instruction execution process in RISC processors required separation of the program and data memories as well as associated busses. This separation is a must to avoid structural hazards in instruction execution pipeline and hence acceleration of the entire computing process [1].

According to Chapter 2, in RCS the control information associated with data processing can be presented in procedural form (e.g., sequence of instructions) and in structural form. This information defines functionality of components, topology of links, and configuration of interfaces in a form of *configuration bit-file* often also called as "configuration bit-stream."

Definition 6.1 *Configuration bit-file* is an information object containing control information for the creation of computing architecture by the determination of all component functionalities, links between components, and procedures of their interaction.

In this case, the RCS should have memory subsystem to store the configuration bit-files associated with all possible configurations of computing and communication resources in FCR.

Obviously, the application may require much more resources than the particular FCR can supply. On the other hand, not all segments of the task may require activation at the same time.

Thus, *configuration memory* reflecting the aforementioned specifics should be divided into several parts:

1. The part containing configuration bit-file that determines current architecture deployed in FCR at all levels. This part of configuration memory is embedded to the PLD architecture because it is directly connected to the on-chip circuitry of the resources deployed in the FCR at on-chip level. This part of configuration memory is similar to the instruction register in instruction-based CPUs conforming CPU circuits to current operation.

2. The part containing configuration bit-files for architectures that may be requested soon. From conceptual point of view, this part of configuration memory can be considered as *configuration cache* because this part of memory is similar in function *to instruction cache* in computers with Harvard architecture [2]. Physically, this type of configuration memory can be located on chip or deployed in external (to PLDs) memory units. In case of on-chip implementation, the configuration cache can be based on internal blocks of SRAM. This way of implementation is possible only for PLDs equipped with self-configuration circuitry of their architecture. Alternatively, onboard-level configuration cache requires special external SRAM-based memory devices and associated configuration management and control units.

3. The part containing configuration bit-files of all possible architectures associated with current workload. This part of memory is similar to the *main memory* of the conventional computer systems containing program files of the active tasks of the workload [2]. The memory storing configuration bit-files of computing architectures conforming active tasks is allocated onboard level or even at system level. This memory can be flash, SRAM, or dynamic RAM (DRAM) based and works under the control of special configuration controllers.

4. The part containing configuration files of all applications that can run on the RCS. This part of configuration memory is similar to the *secondary storage* in conventional computers [3]. The secondary storage of configuration bit-file could be allocated at the system level on external devices (e.g., hard disk drives [HDDs], flash memory devices, remote servers).

The interaction between the on-chip, onboard, and system levels of configuration memory (considering on-chip configuration cache memory) is illustrated in Figure 6.1.

In Figure 6.1, the system FCR consists of several on-chip FSRs deployed in the PLDs: **PLD 1**,..., **PLD N**. The FCR on a system level includes the configurable communication infrastructure often based on configurable communication unit (CCU; discussed in Chapter 5). Each of these devices contains configuration memory directly attached to the configurable on-chip resources. The configuration bit-file, being loaded to the on-chip configuration memory of each **PLD i**, where $i = 1, 2,..., N$, creates architecture of the reconfigurable processing unit—*RPUi* in this PLD. On the other hand, loading the configuration

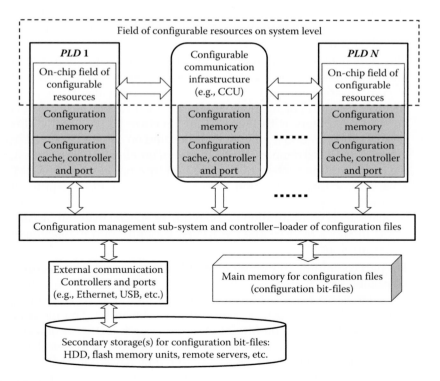

FIGURE 6.1
Generic organization of a configuration memory hierarchy in multilevel RCS.

bit-file to the CCU forms the topology of communication links between all RPUs configured in this system. In turn, the complete architecture of the application-specific processor is programmed at the onboard or system level of the FCR. Hence, in general case, the set of configuration bit-files should be loaded into all configurable logic devices prior to the start of the task execution.

The procedure of loading the configuration bit-stream to the on-chip configuration memory is conducted by an associated on-chip configuration controller via configuration port of the PLD. The on-chip configuration controller could be a dedicated circuit dealing with the external controller-loader via special configuration port of the FPGA or CPLD. For some types of FPGA devices, the on-chip configuration controller can be a custom (configurable on-chip) logic device (soft core) able to program the on-chip configuration memory via internal configuration access port (e.g., Xilinx ICAP [4]). This enables creation of the on-chip cache for configuration bit-files. This aspect will be discussed in Section 7.8.

In contrast to the on-chip cache, the configuration management subsystem with the controller-loader of configuration bit-files is a mandatory part for all classes of RCS.

Definition 6.2 *Controller-loader* is a device directly interfaced to the *configuration memory* of target PLD(s) via configuration port of PLD(s) dealing with the on-chip configuration controller.

The controller-loader on the other hand is interfaced to the main memory for configuration bit-files. The controller-loader can be a commercial off-the-shelf device provided by the FPGA vendor (e.g., [5]) or custom-built device implemented in another PLD (CPLD or FPGA) or microcontroller. The configuration process at on-chip level of RCS will be discussed in detail in the next chapter. Configuration management system being deployed in PLD and/or microcontroller should interact with system ports of the RCS (e.g., USB, Ethernet, SD-card reader) providing communication with secondary storage. This is necessary to update existing set of configuration bit-files in the main configuration memory. The secondary storage can be implemented in any memory device used for the same role in conventional computing systems: HDDs, flash-based memory cards, remote servers, etc.

6.3 Concept of Virtualization of Hardware Resources in the RCS

The hierarchical organization of the configuration memory subsystem in RCS may give the same benefits as virtual memory (VM) organization in conventional computing architectures. The major benefit of the utilization of the VM approach in the instruction-based computers is that it significantly increases the cost-efficiency of computing and memory resources. Indeed, the main idea of having a hierarchical organization of memory in the conventional computing systems is to ensure the loading of only that control and data information to the CPU that are needed just at the time of execution. The information that may be referenced soon should be stored in special memory module located close to the CPU. This storage, called cache, utilizes prediction mechanism based on spatial and temporal locality of information encoded in the program. In other words, it is assumed that if some instruction or data-word has been referenced by CPU, it is quite probable that words located close to referenced words (in address space) may be referenced soon (spatial locality). On the other hand, same instruction or data could be referenced soon again due to the program loops or cycles of the same operations (temporal locality). The earlier probability is high due to the sequential nature of the programming and instruction/data-execution process organization in conventional computers [2]. However, the cost of SRAM-based cache is relatively high in comparison with that of DRAM. Therefore, the utilization of cheaper DRAM as main instruction and data memory can significantly reduce the system cost. Furthermore, the rest of the workload associated with nonactive tasks or task segments can be stored in external (secondary) storages (e.g., HDDs, flash memory cards). The cost of 1-bit storage is cheaper in order(s) of magnitude on the secondary storage when compared to DRAM, and the cost of data storage in DRAM is cheaper compared to that of SRAM-based cache. This allows keeping the task in virtual address area, when some portion of the task instructions and data is loaded in the CPU and is in the process of execution, while some part is stored in the cache waiting execution relatively soon and some task segments (pages) are distributed between the main memory and secondary storages according to the probability of referencing. Keeping most of the task instructions and data in secondary storages allows significant reduction in system memory cost and therefore cost of the entire computing system. On the other hand, the average data access time is somewhat higher than SRAM access time but certainly lower than the respective DRAM access time.

The same idea is applicable to the RCS memory hierarchy organization. As mentioned in Chapter 2, the task segments of the application may not need to be executed on the RCS simultaneously. There can also be multimodal applications when only one of the modes is active at a time. Data and event dependencies may cause certain schedule of task segments initiation. Therefore, all nonactive task segments represented in a form of configuration bit-files can be stored in the main configuration memory or even on secondary storage(s) of the RCS. This fact allows reduction in high-cost computing resource utilization and therefore significant reduction in the system cost. Consequently, keeping nonactive tasks or task segments in the main configuration memory enables minimization of power consumption as well. In addition to that, hierarchical organization of configuration memory makes possible performing of data-execution simultaneously with updating the configuration bit-files in the main configuration memory.

To illustrate the effectiveness of the earlier concept, let us consider the example of multimodal RCS. Let us assume that RCS can work in 1 of the 10 modes at any given period of time. Each mode requires 50% of the resources permanently used in all modes of operation. These resources can be involved in interfacing to system peripherals and memory modules, buffering, and preprocessing input data, clock distribution, and synchronization circuits, etc. The rest 50% of the resources need to be reconfigured to accommodate the next mode. To simplify the example, we will consider statically reconfigurable RCS. This class of RCS was discussed in Section 2.4. In this type of reconfigurable systems, mode switching process requires complete reconfiguration of resources in the FCR. Therefore, RCS should store in its main configuration memory 10 separate configuration bit-files, each of which is associated with the respective mode of operation. The block diagram of this RCS is shown in Figure 6.2. In start-up moment, the initial mode is activated by loading one of the initial configuration bit-files (e.g., configuration bit-file for mode M_1). When mode should be changed (according to external or internal events), the controller-loader sends the start address of the requested configuration bit-file associated with this mode (e.g., M_i) to the main configuration memory. Then, the loading process starts by reading word by word this bit-file and writing these words to the configuration port of the target FPGA device. The internal configuration controller then translates the configuration data to the on-chip configuration memory according to the internal addresses of on-chip configuration memory cells. At the end of the process, the on-chip controller verifies the check data of the received

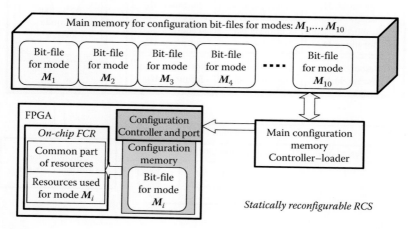

FIGURE 6.2
Block diagram of the multimodal statically reconfigurable computing system.

configuration bit-file with accumulated check data and, if all is correct, activates the processing circuits in the FCR. Thus, each mode of operation requires certain period of reconfiguration and further global reset before the start of a new mode.

As mentioned earlier, half of the resources (50%) are common part for all modes of operation and the rest are custom for each mode reflecting mode-specific functionalities.

Let us take as an example the Xilinx Virtex-6 FPGA family and assume that the processing architecture compiled for this family of devices requires up to *20,000 CLB-based logic slices and up to 512 kB block RAM for each mode of operation*. Hence, the amount of resources commonly used for all modes is equal to *50% × 20,000 slices = 10,000 logic slices and 50% × 512 kB = 256 kB* of embedded block RAM. Therefore, the reconfigurable part for each of the 10 modes will require the same amount of resources as for the common part.

Let us now consider the conventional FPGA-based system where FPGA accommodates all circuits associated with all modes of operation in one design. In this case, the FPGA is keeping the *complete hardware design* in on-chip configuration memory. The block diagram of this ASIC-like FPGA-based system is depicted in Figure 6.3.

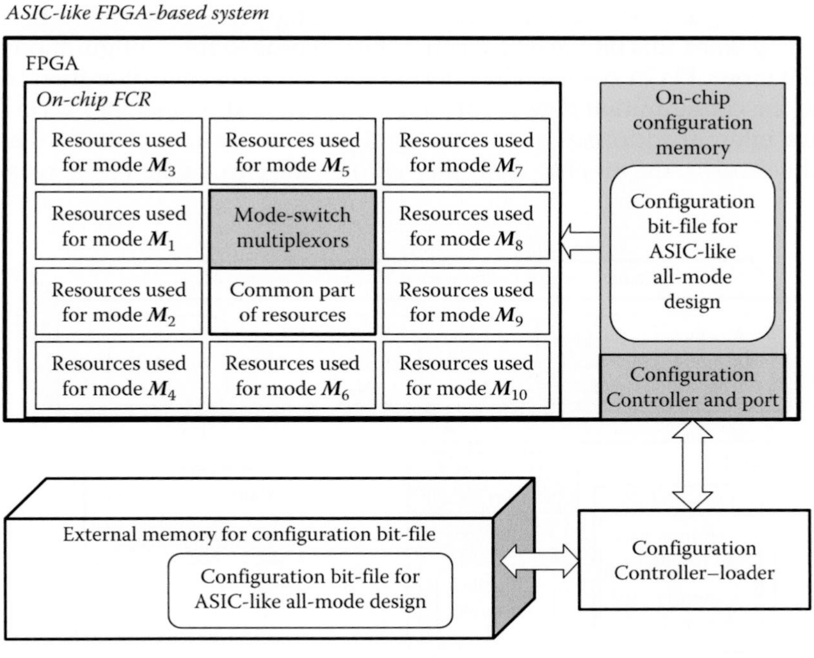

ASIC-like FPGA-based system

FIGURE 6.3
Block diagram of an application-specific integrated circuit–like field-programmable gate array–based computing system.

This system will require the following amount of on-chip logic and memory resources:

$$\text{Logic resources} = 10,000 \text{ CLB slices (common part)} + 10 \text{ (modes)} \times 50\% \\ \times 20,000 \text{ CLB slices (mode-specific parts)} = 10,000 \text{ CLB slices} + 10 \\ \times 10,000 \text{ CLB slices} = 110,000 \text{ CLB slices}$$

$$\text{Memory resources} = 256 \text{ kB (common part)} + 10 \text{ (modes)} \\ \times 50\% \times 256 \text{ kB (mode-specific parts)} = 2816 \text{ kB}$$

Note: The *mode-switch multiplexors* and additional support circuits are excluded from the resource count to simplify the comparative analysis.

Taking the same Xilinx Virtex-6 family of FPGAs as an example, it is possible to find out the most suitable FPGA devices for the aforementioned solutions:

1. *XC6VLX130T* with 20,000 CLB slices and 1,188 kB of block RAM is sufficient for the implementation of the statically reconfigurable RCS.
   The cost of this device is in the range of ~U.S. $1000.
2. *XC6VLX760* with 118,560 CLB slices and 3,240 kB of block RAM is the only FPGA device (in this family) sufficient for utilization in ASIC-like FPGA-based computing system.

The cost of this device is in the range of ~U.S. $20,000.

Thus, the cost of the FPGA device utilized for ASIC-like system is approximately 20 times higher than the cost of the FPGA device sufficient for statically reconfigurable RCS.

The reason of this significant difference of FPGA costs is quite clear. In ASIC-like FPGA system, all functionalities associated with all modes of operation are "stored" in the form of *actual hardware* circuits in the FPGA. However, only few of the aforementioned functionalities are actively used in each mode of operations. The rest of the expensive computing and memory resources included into all-modes architecture in the FPGA are in idle stages just waiting for their modes. In contrast to that, the statically reconfigurable RCS solution allows keeping nonactive functionalities in much cheaper memory resources until these resources are requested for operation. In other words, these functionalities are kept in the form of *virtual hardware*—configuration bit-files.

Certainly, there is some hardware overhead associated with additional devices such as controller-loader and larger memory unit to store 10 configuration bit-files (Figure 6.2) instead of one configuration bit-file for the ASIC-like FPGA-based system (Figure 6.3).

Let us estimate the approximate costs of this hardware overhead.

The controller-loader can be implemented on the base of embedded micro-controller, CPLD, or small FPGA. In general case, this device should perform the following functions:

1. Receive the code of the requested mode.
2. Generate the initial address of the configuration bit-file for the main configuration memory.
3. Conduct loading procedure to the on-chip configuration memory of the target FPGA according to the configuration port standard protocol.

In all variants of implementation, the cost of controller-loader will be in range of U.S. \$10–\$40. For example, if the controller-loader is implemented on the Xilinx Spartan-3AN FPGA [6] with an embedded nonvolatile configuration memory (e.g., XC3S50AN-4TQG144C), the unit cost will not exceed U.S. \$10. This cost can be considered as negligible in comparison with the cost of FPGA selected for data-processing. The same is true for the additional memory cost used for larger main configuration memory. As per Xilinx Virtex-6 FPGAs overview [7], the configuration bit-files for the complete FPGA configuration range from 26 Mb (3.25 MB) to 177 Mb (22.2 MB). Even in case when 10 largest configuration bit-files for 10 separate modes of operation have to be stored in external flash memory, the total memory volume will not exceed 10×22.2 MB $= 222$ MB. The cost of the closest NAND flash memory IC with 256 MB (e.g., Micron Technology Inc. MT29F2G08ABAEAWP:E TR) would range from U.S. \$5 to \$10 per device.

In other words, the additional cost of the hardware overhead including configuration controller-loader and external configuration memory will not exceed U.S. \$20–\$50.

Therefore, the total cost of hardware components of statically reconfigurable RCS can be estimated (with certain approximation) excluding the cost of the PCB, power supplies, cooling, and other subsystems as follows:

$$\text{Cost(SRCS)} = \text{Cost of FPGA} + \text{Cost of configuration controllers and memory}$$

$$\simeq \$1000 + \$50 = \sim\mathbf{\$1050}$$

In contrast to this RCS, the cost of components in ASIC-like FPGA-based system is close to

$$\text{Cost (ASIC-FPGA)} = \text{Cost of FPGA devices}$$

$$+ \text{Cost of configuration controllers and memory}$$

$$\simeq \$20,000 + \$20 = \sim\mathbf{\$20,020}$$

The cost-efficiency between these two systems can be estimated by ratio:

$$Rrcs/fpga = \text{Cost(ASIC-FPGA)}/\text{Cost(SRCS)} = \$20{,}020/\$1{,}050 = \mathbf{19}$$

This ratio shows that the implementation of this application in the form of statically reconfigurable RCS costs 19 times less than the implementation in the form of large FPGA-based system.

Certainly, many factors are not considered in this estimation. As was mentioned earlier, the cost of PCB design, layout, and production plus the cost of power supplies, convertors, interface elements, cooling system, etc. is not taken in consideration. Only the components associated with data-processing and architecture reconfiguration have been compared with each other to illustrate potential reduction in the hardware cost of these components.

For more detail, this aspect will be discussed and analyzed in the following chapters.

6.4 Summary

In sum, it is possible to say that the representation of the processing and communication hardware resources and therefore SoC architecture in *virtual form* as a set of configuration bit-files enables reduction in the system cost in orders of magnitude. Furthermore, the utilization of smaller reconfigurable logic device reduces the cost and power consumption of the system without sacrificing the performance characteristics.

On the other hand, there is one aspect that can be an important factor in the selection between the ASIC-like system with static hardware architecture and the statically reconfigurable RCS. This aspect is associated with switching time between modes of operation and respective system functionalities. If this time should range from nanoseconds to units of microseconds, the ASIC-like system design solution is the only way for SoC implementation. In this case, the mode switching process is based on set of on-chip multiplexors. These multiplexors can form the required processing architecture by switching digital circuits already configured in FCR. This process can be performed in range of nanoseconds. If this is not a case and application allows changing the system functionalities that range from hundreds of microseconds to milliseconds, then the reconfiguration of the FCR can be considered as a more efficient solution.

Therefore, the organization of reconfiguration process becomes the most important aspect in the RCS architecture organization. The proper organization of the reconfiguration process directly influences on the system organization of the RCS and its cost-efficiency. Thus, changing the system functionality should be considered in all levels of the FCR: on-chip, onboard, and system levels. Hence, the architecture configuration and reconfiguration processes should be considered in each of these levels because each level of the RCS has its own specifics.

Exercises and Problems

Exercises

1. Why does RCS need to have the configuration memory in its architecture in addition or in replacement of the program memory?
2. What is the main reason that dictates configuration memory hierarchy in RCS architecture?
3. Determine all levels of configuration memory hierarchy and define the functions of each level of configuration memory hierarchy.
4. What is the role and purpose of controller-loader of configuration bit-files?
5. What are the ways of implementation of the controller-loader in the RCS architecture?
6. Why do configuration management subsystems need to be interfaced to the secondary storage(s) of configuration bit-files and/or external sources of configuration information?
7. Describe the main idea for the virtualization of hardware resources using virtual memory as an example. Extend this concept on data-processing and communication resources.
8. What are the main reason and benefits of the virtualization of hardware resources in RCS? What are the drawbacks?

Problems

Using the example discussed in Section 6.3 and illustrated in Figures 6.2 and 6.3, analyze the effectiveness of statically reconfigurable RCS in comparison with ASIC-like FPGA system if the following applies:

1. The numbers of modes of operation (variations of algorithm and/or data structure) are as follows: (a) 20, (b) 30, and (c) 40 modes.
   Assuming the same family of FPGA devices and the possibility to implement the ASIC-like system in a set of 2, 3, or more FPGAs, determine the cost-efficiency of RCS in comparison with ASIC-like FPGA system according to the number of modes.
2. The number of modes of operation is 16 and the common part of the SoC for each of the 16 modes requires 5000 CLB slices and 128 kB of embedded memory resources. However, the custom part of the logic resources for each mode of operation is approximately equal to 14,000

CLB slices and 360 kB memory blocks. Estimate the cost of ASIC-like FPGA-based system and compare it to the cost of statically reconfigurable RCS assuming that more than one FPGA can be used in the ASIC-like system.

Note: Consider the cost of the FPGA devices and the cost of supporting circuits (memory and controller) same as in the presented example.

References

1. J. L. Hennessy and D. A. Patterson, Pipelining: Basic and intermediate concepts, in *Computer Architecture: A Quantitative Approach*, 3rd ed., Morgan Kaufmann Publishers, San Francisco, CA, 2003, Appendix-A, pp. A1–A87.
2. J. L. Hennessy and D. A. Patterson, Memory Hierarchy Design, in *Computer Architecture: A Quantitative Approach*, 3rd ed., Morgan Kaufmann Publishers, San Francisco, CA, 2003, Chapter 5, pp. 390–504.
3. J. L. Hennessy and D. A. Patterson, Storage Systems, *Computer Architecture: A Quantitative Approach*, 3rd ed., Morgan Kaufmann Publishers, San Francisco, CA, 2003, Chapter 7, pp. 678–770.
4. J. Ayer Jr., Dual use of ICAP with SEM controller, Xilinx Application Note: Virtex-6 and Spartan-6 devices, XAPP517 (v1.0), Xilinx, Inc., San Jose, CA, December 2, 2011.
5. Xilinx, *Platform Flash PROM*, User guide, UG161 (v1.5), Xilinx, Inc., San Jose, CA, October 29, 2009.
6. Xilinx, Spartan-3AN FPGA family data sheet, DS557, June 12, 2014.
7. Xilinx, Virtex-6 family overview, Product specification, DS150 (v2.4), Xilinx, Inc., San Jose, CA, January 19, 2012.

References

7

Reconfiguration Process Organization in the On-Chip Level of a Reconfigurable Computing System

7.1 Introduction

In Chapter 6, the generic organization of configuration memory hierarchy was considered. The concept of hardware virtualization was discussed in accordance with the aforementioned memory hierarchy. The cost-efficiency of hardware virtualization was considered for multimodal applications. It was shown that significant reduction in system cost by virtualization of hardware circuits is possible because in multimodal applications only a part of system functionalities is active at a mode. The rest of the functionalities could be stored in the form of configuration bit-files in the external memory. The drawback of this approach is the extended mode-switching time associated with the time for the reconfiguration of resources at the on-chip and system level of a reconfigurable computing system (RCS). Therefore, the organization of the reconfiguration process plays a vital role in the entire architecture organization of RCS. The organization of reconfiguration process can be divided into two parts: reconfiguration of resources at the on-chip level of the field of configurable resource (FCR) and the system level of the FCR reconfiguration.

This chapter is dedicated to the organization of static configuration and dynamic reconfiguration processes at the on-chip level of the RCS. This assumes organization of (1) on-chip configuration memory programming and its partitioning for configuration frames and reconfigurable partitions (RPs), (2) consideration of the types of configuration ports and associated buses, (3) on-chip level configuration cache memory organization, and (4) organization of on-chip configuration controller(s) and their interaction with associated port(s) and memory modules at the on-chip and system levels of the RCS architecture.

7.2 Reconfiguration of the On-Chip Resources in the RCS

As it was mentioned in Section 6.2, the configuration of the on-chip architecture in the FCR assumes loading the configuration bit-file(s) to the on-
chip configuration memory of the programmable logic device (PLD) (e.g.,
field-programmable gate array [FPGA], complex programmable logic device
[CPLD], or coarse grained reconfigurable array [CGRA]). Thus, the period of
on-chip architecture configuration in the FCR—$T_{fcr\text{-}conf}$—depends on (1) the
volume of configuration bit-file for the on-chip FCR (V_{fcr}) and (2) the configuration port and channel bandwidth (BW_{conf}).
   Thus,

$$T_{fcr\text{-}conf} = \frac{V_{fcr}}{BW_{conf}} \qquad (7.1)$$

Hence, a larger volume of the bit-file and/or lower configuration channel
bandwidth causes longer (re)configuration periods and vice versa. In other
words, the more the resources that need to be reconfigured to accommodate
new functionalities, the more time it will take. Obviously, complete reconfiguration of all resources in the FCR will take maximum reconfiguration
time for this on-chip FCR. Changing the functionalities of the system often
means reaction to changes in the environment or internal requests. In many
practical cases, this reaction time is a vital parameter that must not exceed
the given limit. Thus, in general case, switching time between configurations of FCR is one of the critical characteristics for RCS. Certainly, complete
reconfiguration of the on-chip resources is the simplest process in comparison with other approaches. However, reconfiguration time for the PLD may
exceed the required time constraints. In addition, the resource utilization
factor (discussed in Chapter 2) can decrease due to longer reconfiguration
time when resources being in reconfiguration process cannot be used for
data-execution. The alternative solution that accelerates the mode-switching
process is dynamic partial reconfiguration. This method of FCR reconfiguration has been discussed in Sections 2.4 and 2.5. A more detailed information
on this approach will be considered in the following sections.

7.3 Partitioning the On-Chip Field of Configurable Resources

As per definition of partially reconfigurable FCR (Definition 2.11) given in
Chapter 2, this method of resource reconfiguration allows changing the functionality of some parts of resources in the FCR when the rest of the resources
keep the same functionalities. Obviously, reduction in the amount of resources

to be reconfigured, respectively, reduces their reconfiguration time. Hence, if only a few functions need to be updated from mode-to-mode transition, the mode-switching time can be significantly lowered by reconfiguration only that part of the resources that are necessary to accommodate the new functionalities.

Definition 7.1 *Partially reconfigurable region* (PRR) is the part of the FCR where resources can be reconfigured independently from the rest of the resources deployed in the FCR.

Note: It is assumed that the boundaries of the PRR in the FCR are not fixed by any structural constraints but can be defined by an on-chip partitioning procedure.

Definition 7.2 *Base configurable region* (BCR) is the smallest configurable region of resources structurally determined in the FCR.

As per the earlier definition, the BCRs are considered as building blocks for any PRRs in the on-chip FCRs. For example, in Xilinx Virtex families of FPGA devices, the BCRs are called base regions [1] and are associated with *configuration frames*. For example, BCR in Xilinx Virtex-6 FPGA devices are organized as rectangular on- chip region (column) 1 CLB wide by 40 CLBs high. The BCR in Xilinx Virtex-5 is twice smaller column which consist of 20×1 CLBs = 20 CLBs [1].

Thus, the rectangular area of the on-chip FCR can be divided into a certain number of BCRs. For example, for the Xilinx Virtex-6 FPGA devices mentioned in Section 6.3, there are 16,840 BCRs for XC6VLX130T and 71,262 BCRs for XC6VLX760.

7.4 Partitioning the On-Chip Configuration Memory for Partial Reconfiguration

The on-chip configuration static random access memory (SRAM) is usually a 1-bit organized addressable memory unit. The 1-bit organization is dictated by a fine-grained organization of a majority of logic and routing resources in the on-chip FCR. It would be difficult to address every single bit in the on-chip configuration memory because it significantly increases the address area of this memory. Hence, 1-bit-wide configuration memory cells should be grouped together to be addressed then as the multibit configuration word or sequence of words. In general case, any part of the resources grouped by respective configuration bits (e.g., LUTs for CLBs, routing pass transistors) can be addressable portions reflected in the aforementioned multibit configuration words. Therefore, the organization of the configuration memory should conform to the organization of the partitioned on-chip FCR. Since the smallest configurable region in the on-chip FCR is the BCR, it make sense to combine all configuration bits associated with resources allocated in the BCR to the smallest addressable portion of the on-chip configuration memory.

Definition 7.3 *Configuration frame* is the smallest directly addressable segment of the *on-chip configuration memory* associated with the smallest configurable region of hardware resources in the on-chip FCR.

For example, in Xilinx Virtex-6 family of FPGA devices, each configuration frame consists of 2592 bits of configuration information organized as 81 words $\times$ 32 bits per word [2]. Each of these configuration frames in the on-chip configuration memory is directly associated with respective BCR consisting of 40×1 CLB slice. The correspondence between the resources in BCR and associated configuration frame in the on-chip configuration memory is illustrated in Figure 7.1.

The entire on-chip configuration memory for partially reconfigurable FCR can be organized as 2D array of configuration frames, each of which controls the functionality of the associated resources. Since heterogeneous on-chip FCR consists of different types of processing, routing, and memory resources, the configuration frames reflecting control specifics of these resources should be resource specific. In other words, the configuration frames should be divided into frames associated with logic resources (e.g., CLB slices), memory resources (e.g., block RAMs), I/O resources, etc. However, the configuration frames is expected to be the same in length. The equal length of the configuration frames regardless of the types of resources

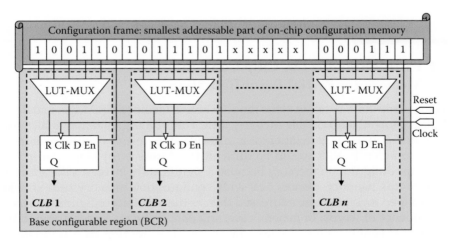

FIGURE 7.1
Correspondence between the resources in the BCR and the associated configuration frame.

Configuration frame #1:1	Configuration frame #1:2	Configuration frame #1:3	Configuration frame #1:i	Configuration frame #1:K
Configuration frame #2:1	Configuration frame #2:2	Configuration frame #2:3	Configuration frame #2:i	Configuration frame #2:K
Configuration frame #3:1	Configuration frame #3:2	Configuration frame #3:3	Configuration frame #3:i	Configuration frame #3:K
Configuration frame #j:1	Configuration frame #j:2	Configuration frame #j:3	Configuration frame #j:i	Configuration frame #j:K
Configuration frame #S:1	Configuration frame #S:2	Configuration frame #S:3	Configuration frame #S:i	Configuration frame #S:K
Configuration memory for: synchronization circuits, global and local clock generation, clock distribution, PLLs/DLLs, internal configuration ports, etc.				

FIGURE 7.2

General organization of the configuration SRAM for the partially reconfigurable FCR.

to be reconfigured significantly simplifies the composition of configuration bit-files. Furthermore, bit-file generation, complexity of on-chip configuration controller, and many other aspects associated with partial reconfiguration of the on-chip FCR also will be more simple than in the case of resource-specific frames. Therefore, the organization of the on-chip configuration memory will look like 2D-array of identical segments of configuration SRAM (configuration frames) identified by individual addresses. The general organization of such configuration memory is presented in Figure 7.2.

Configuration frames should be distributed on-chip as close as possible to the respective base regions (BCRs). This minimizes the length of control lines from configuration memory cells to the logic, routing, data memory, or I/O circuits. Thus, the address of configuration frames can be divided into row and column parts.

In sum, configuration frame can be considered as the control part of the column-organized, multibit processing element (PE), I/O element (IOE), or distributed memory element (ME). These multibit processing, interface, or MEs can be configured for a certain mode of operation by loading the partial configuration bit-file to the addressable configuration frames, and the address of this frame is associated with the placement of the respective BCR.

The configuration of more complex function-specific PE in the selected PRR requires loading of several bit-files to the respective configuration frames associated with BCRs allocated in this PRR. For example, if PRR is bordered by rows 1 and 3 and columns 2 and 4 (shown in Figure 7.3 by bold-edge rectangle), it consists of configuration frames #1:2, #1:3, #2:2, and #2:3. Hence, the configuration of this PRR can be done by loading four configuration bit-files to the configuration frames listed earlier.

As of the independent nature of configuration frames, in general case, PRR of any "shape" could be created from these "bricks." However, the complete configuration of the entire on-chip FCR cannot be performed only by

Configuration frame #1:1	Configuration frame #1:2	Configuration frame #1:3	Configuration frame #1:i	Configuration frame #1:K
Configuration frame #2:1	Configuration frame #2:2	Configuration frame #2:3	Configuration frame #2:i	Configuration frame #2:K
Configuration frame #3:1	Configuration frame #3:2	Configuration frame #3:3	Configuration frame #3:i	Configuration frame #3:K
Configuration frame #j:1	Configuration frame #j:2	Configuration frame #j:3	Configuration frame #j:i	Configuration frame #j:K
Configuration frame #S:1	Configuration frame #S:2	Configuration frame #S:3	Configuration frame #S:i	Configuration frame #S:K
Configuration memory for: synchronization circuits, global and local clock generation, clock distribution, PLLs/DLLs, internal configuration ports, etc.				

FIGURE 7.3
Configuration of the partially reconfigurable region in the on-chip FCR.

configuration of all BCRs. There are several regions of resources that are common for all PEs, IOEs, and MEs. These are synchronization, global reset, global and local clock generation, distribution and manipulation circuits, and devices such as PLLs or DLLs and internal configuration ports (if any). These circuits should be included in the static part of the on-chip configuration. This is needed because these circuits should support and synchronize the rest of the resources in the on-chip FCR continuing their operations while reconfiguration process of other resources is ongoing. The static resources configuration should be done at start-up time. Therefore, initially, the complete configuration of the on-chip FCR is a must. During this initial complete configuration of the on-chip FCR, all circuits common for all modes of operations should be configured. At the same time, some areas of configuration memory could be reserved for further reconfiguration of the respective PRR.

Definition 7.4 *Reconfigurable partition* in the on-chip configuration memory is the determined segment of this memory *consisting only from configuration frames* associated with the PRR to be reused by different segments of the application(s).

Definition 7.5 *Virtual hardware component* (VHC) is an information object that is (a) dedicated to perform the determined set of functions and (b) represented by partial configuration bit-file to be downloaded to the assigned RP of the on-chip configuration memory.

As per this definition, there could be many VHCs targeting the same RP. These VHCs should be loaded to the RP according to the mode of operation and only for the period of operation associated with this mode. Otherwise, they are located in the main configuration memory or even secondary storage to minimize the hardware and power utilization. As it was shown in Section 6.3, it enables significant reduction on the cost of the RCS and its power consumption.

On the other hand, there could be many VHCs performing the same set of functions but targeting different RPs. The main reason for this is the non-identical floor mapping of resources in different PRRs. In many practical cases in partially reconfigurable FPGAs, VHC designed for RP_i may not work at RP_j or even jeopardize VHCs allocated in neighboring RPs.

In the case of multimodal operations, each mode requires some static part of resources common for all modes. Thus, the complete initial configuration of the on-chip FCR is needed. This start-up configuration is setting up all *static logic* or so-called top logic that includes on-chip clock generation, clock distribution, synchronization circuits, and configuration frames associated with BCRs responsible for operations common for all modes (e.g., interface to peripherals and off-chip memory modules, control and preprocessing operations). The area configured by initial start-up complete configuration procedure is shown in gray in Figure 7.4.

In this configuration, some reserved RPs, that is, RP_1, RP_2, and RP_3, are left blank. Since RPs are address areas in the on-chip configuration memory, it means that there are no data loaded to the associated configuration frames (e.g., zeros or FFs). Thus, the BCRs controlled by the aforementioned configuration frames will not perform any logic functions or create any routing topologies. These RPs are shown as white areas in Figure 7.4. Then, RPs could be used for sets of VHCs customizing on-chip architecture to the requested mode of operation. These VHCs can be configured in the RPs by loading series of partial bit-files. For example, in mode M_i, VHC5 should be loaded to the RP_1 and VHC2 to the RP_3. RP_2 is left blank, and in mode M_j, VHC7 should be placed to the RP_1 and VHC8 to the RP_2, and VHC2 continues to stay in the RP_3.

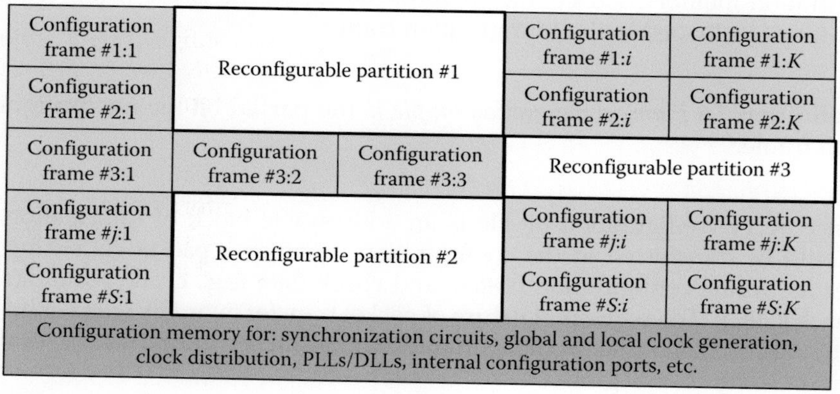

Configuration frame #1:1	Reconfigurable partition #1		Configuration frame #1:i	Configuration frame #1:K
Configuration frame #2:1			Configuration frame #2:i	Configuration frame #2:K
Configuration frame #3:1	Configuration frame #3:2	Configuration frame #3:3	Reconfigurable partition #3	
Configuration frame #j:1	Reconfigurable partition #2		Configuration frame #j:i	Configuration frame #j:K
Configuration frame #S:1			Configuration frame #S:i	Configuration frame #S:K
Configuration memory for: synchronization circuits, global and local clock generation, clock distribution, PLLs/DLLs, internal configuration ports, etc.				

FIGURE 7.4
Initial complete configuration of the on-chip configuration memory with reserved (blank) reconfigurable partitions (RPs): RP_1, RP_2, and RP_3.

Therefore, only at start-up time the complete configuration of the on-chip FCR is needed. All further mode switching requires loading only partial bit-files associated with VHCs requested for the upcoming mode. In turn, this allows minimization of

1. Mode-switching time or external event response time
2. On-chip configuration memory overhead caused by RPs under reconfiguration
3. Hardware and power consumption of the entire FCR

All these became possible because most of the functionalities represented in VHCs are stored in the off-chip memory units but not in the form of active on-chip resources. It is also necessary to mention that the boundaries or RPs are not static and can be changed dynamically. However, in this case, respective reconfiguration of the static part of the architecture may be needed.

7.5 Reconfiguration Process and Configuration Bit-File Structure

As per the definition of RP, the configuration of RP assumes configuration of all frames CF_i, $i = 1, 2,\ldots, m$ included in this RP. In other words, the configuration of the RP requires loading partial configuration bit-files to all configuration frames composing this RP. Thus, the organization of partial bit-file to be loaded to the configuration frame should be considered first. As was mentioned earlier, configuration frame is the *smallest addressable part* of a configuration memory. Hence, the smallest partial bit-file is the configuration bit-file to be loaded to the configuration frame.

Definition 7.6 *Frame configuration bit-file* is the partial bit-file for configuration frame.

Since every configuration bit-file is an addressable block of information, it should be transferred to the configuration frame as a packet consisting of header, frame configuration bit-file, and check data (e.g., cyclic redundancy check [CRC]). The sequential nature of packet transfer (word by word or bit by bit) requires some synchronization information. Generic organization of this packet is shown in Figure 7.5.

Initial synchronization data (synchronization byte or word) are necessary for the separation of the new packet from the previous one. Thus, it shows to the receiver (on-chip configuration controller) that the header of the new

Synchronization data	Header (Address)	Frame configuration bit-file	Check data (CRC)

FIGURE 7.5
General organization of frame configuration bit-file packet.

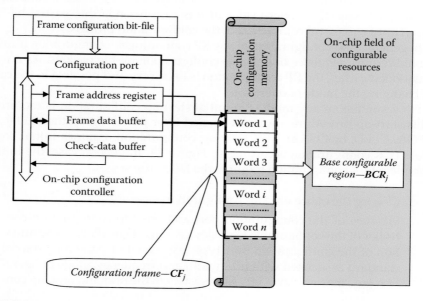

FIGURE 7.6
Configuration of base configurable region by loading partial configuration bit-file.

packet is coming next. The organization of the process of loading the frame configuration bit-file to the configuration frame is shown in Figure 7.6.

The header must contain the initial address of the configuration frame that should be stored in a Frame Address Register (FAR)—a part of the on-chip configuration controller. Taking into account that the size of frame configuration bit-file is the same for any configuration frame, the on-chip configuration controller should have the frame address counter preset on the length of this bit-file. This counter receives the initial address of the configuration frame from the packet header and then increments addresses for each following configuration word. Configuration word to be loaded to the configuration frame is received from the frame configuration bit-file packet and stored in the frame data buffer. Then, this word is transferred to the memory cell of configuration frame addressed by FAR. This process continues until the content of the FAR reaches the limit equal to the number of words in the frame configuration bit-file.

During the aforementioned loading process of frame configuration bit-file, the on-chip configuration controller performs the calculation of check data associated with the frame configuration bit-file (e.g., CRC or check sum). These data are stored in the check-data buffer (CDB). The last part of the frame configuration bit-file packet is the check-data word (e.g., CRC). This check-data word is compared with the calculated check data in CDB. If the check data in the packet is equal to the word in the CDB, the configuration of the respective configuration frame is correct. Otherwise, the on-chip configuration controller sends the negative feedback to the external (onboard) controller-loader. Controller-loader in this case resends this partial bit-file again until successful completion of the configuration process.

In general case, configuration of any RP containing m-configuration frames would require sequence of m-frame configuration bit-file packets. As it was mentioned earlier, the RP configuration is necessary and possible in the case of changing the mode of operation.

However, before any mode, the initial start-up configuration is required. This start-up procedure always is needed after powering-up the PLD (e.g., FPGA, CPLD, or CGRA). This start-up configuration assumes initial complete FCR configuration according to Figure 7.4. In general case, the initial complete FCR configuration requires the following stages:

1. *Clearing* the entire on-chip configuration memory.

2. *Initiation* of the configuration port and on-chip configuration controller of the configurable logic device. This stage allows determination of the configuration port type (e.g., serial or parallel), protocol standard associated with port width and frequency, mode of operation (e.g., master/slave), etc. All registers in the port and on-chip configuration controller should also be reset at this stage.

3. The *synchronization* stage is necessary to determine the start of the configuration bit-files to be loaded to the configuration memory. In other words, any data received before identifying the synchronization word should be ignored by the configuration controller. This synchronization word usually is fixed by the vendor of the reconfigurable device and thus is the same for all other devices provided by this vendor.

4. The *configuration* stage follows the synchronization stage and consists of a series of configuration bit-files for static logic (e.g., global and local clock generation and distribution, PLLs/DLLs) and frame configuration bit-files. During this stage, the check data are calculated and compared with the check data included in configuration bit-file packets.

5. The *start-up* stage initiates the resources in the on-chip FCR after configuration if the aforementioned complete configuration process was successful.

The aforementioned process of complete (full) configuration is illustrated in Figure 7.7. This process can be divided into three sequential parts, each of which requires different periods of time:

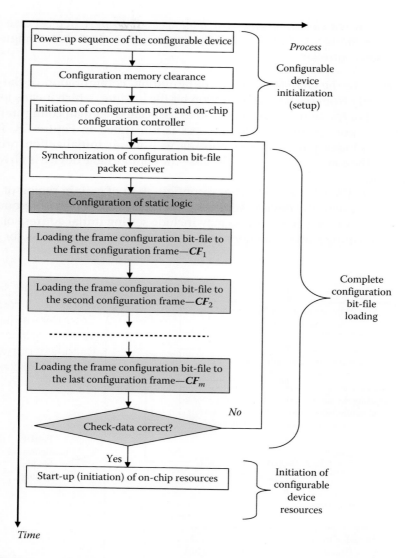

FIGURE 7.7
Process of complete configuration of the on-chip FCR.

1. *Configurable device initiation*: This part of the configuration process takes a relatively short time and mostly depends on the period of power-up sequence and organization of configuration memory clearance.

2. *Complete device configuration*: The timing of this part depends on the organization of configuration port, on-chip controller-loader, and configuration bus bandwidth.

3. *Start-up of architecture configured in the device*: This part, in most cases, is the shortest and depends on the organization of on-chip configuration controller.

In contrast to the complete device configuration, configuration of the RP consists only from partial configuration bit-file loading, which means loading the set of frame configuration bit-files associated with all configuration frames composing this RP. This process is illustrated in Figure 7.8. This process excludes power-up sequence as well as all initialization procedures for the configurable device because the configurable device is already initialized.

Indeed, at the time when partial reconfiguration of certain region of the FCR is requested, the device is initiated and the rest of on-chip resources are already running. The process may include setting initial address of the RP in the configuration memory and autoincrement this address for the

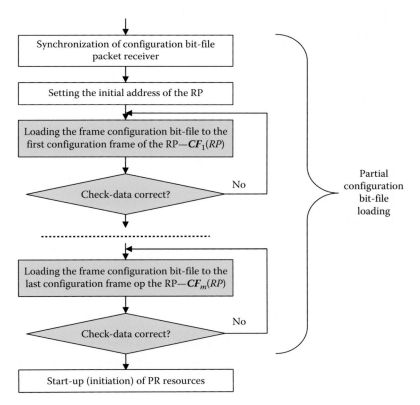

FIGURE 7.8
Process of partial configuration of the reconfigurable partition in the on-chip FCR.

next configuration frame in sequence. Alternatively, each frame configuration bit-file packet should have its own initial address and check data as shown in Figure 7.5.

In addition to that, in the end of the RP configuration process, the RP resources should be initiated. This step, however, may not be the same as it is in the case of complete device reconfiguration. It is necessary to mention that the rest of the resources in the FCR of the device are *already executing their task(s)*. Thus, the initiation of resources associated with RP can be considered as run-time integration of new functional units (e.g., PEs and MEs.) to the rest of active circuits in the FCR. In turn, this means that the aforementioned integration process must avoid any multisourcing and desynchronization of other functional components in the FCR. This aspect will be considered in detail in Chapter 10.

According to the earlier discussion, partial reconfiguration of any RP may take much less time than complete device configuration due to

1. Reduced number of frame configuration bit-file packets to be loaded to the on-chip configuration memory and also reduction in the configuration overhead (configuration commands)
2. Lack of power-up procedure that may be quite long due to the dependency on different external parameters (e.g., power supply inertia, DC–DC convertors, low-pass filters)

Therefore, the period of RP reconfiguration is determined by the number of frame configuration packets, packet volume (in bits), and configuration channel bandwidth (in bit/s).

In sum, the utilization of the partial reconfiguration can significantly accelerate the mode-switching process and therefore reduce the workload or environment adaptation period.

7.6 Relationship between Reconfiguration Time and Organization of Configuration Port and Bus

As discussed in the previous section and illustrated in Figure 7.6, the configuration bit-files are coming to the on-chip configuration memory over the on-chip configuration controller. The on-chip configuration controller receives these bit-files from the external main memory of configuration bit-files via onboard configuration controller-loader, configuration bus, and associated configuration port as shown in Figure 7.9.

Hence, the reconfiguration time directly depends on the configuration bus and configuration port bandwidth—BW_{conf}. This bandwidth, on the other

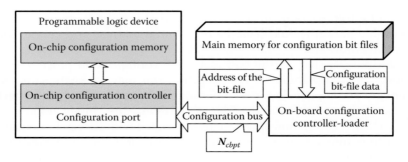

FIGURE 7.9
Configuration controllers' interaction via configuration bus and port.

hand, depends on the width of the configuration bus and port as well as on the bus/port frequency:

$$BW_{conf} = N_{cbpt} * F_{conf\_bus} \tag{7.2}$$

where
N_{cbpt} is the number of configuration bits per transaction
F_{conf_bus} is the configuration bus/port frequency

For example, if the configuration bus is parallel and consist of eight data-lines as well as configuration port of associated PLD, the *Ncbpt* = 1 byte/clock cycle. In case when configuration bus frequency *Fconf_bus* = 50 MHz than

$$BW_{conf} = 50,000,000 \text{ clock cycles/s} \times 1 \text{ byte/clock cycle}$$
$$= 50,000,000 \text{ byte/s}$$
$$= 50,000,000/(1,024 \times 1,024) = \textbf{47.684 MB/s}$$

Note: 1 kB = 1,024 bytes and 1 MB = 1,024 × 1,024 = 1,048,576 bytes.

As any other types of buses, the configuration bus could be serial or parallel, synchronous or asynchronous, or shared or peer to peer. For all known practical cases associated with SRAM-based or flash memory–based on-chip configuration memories, the configuration buses are organized as synchronous serial or synchronous parallel buses. However, in general case, there is no limitation to utilize other types of bus or configuration port organizations.

The serial bus and port consists of one data-in line and one data-out line synchronized by configuration clock signal line. Some additional control lines could be used as well in these types of ports and buses (e.g., master/ slave selector, busy/done handshake signals).

The most common standards used for the serial configuration buses and ports are (1) the Standard Test Access Port and Boundary-Scan

Architecture—JTAG bus (IEEE Standard 1149.1 [3])—and (2) serial peripheral interface (SPI) [4]. The details associated with the actual implementation of the aforementioned configuration ports and buses are provided in the vendor's documentation (e.g., JTAG configuration for Xilinx Virtex-6 FPGA [5] or for ALTERA FPGA [6]).

The parallel synchronous configuration bus consists of a number of bidirectional I/O data-lines and synchronization clock signal line as well as additional control/mode-selection lines. These parallel buses are usually 8/16- or 32-bit word organized (e.g., SelectMAP-8/16/32 in Xilinx Virtex FPGA devices [7] or FPP-8/16/32 in Altera Stratix V FPGA [8]).

Knowing the configuration bus/port bandwidth, the RP configuration time (T_{RPi_conf}) can be determined according to the m_i-number of configuration frames in the RP (RP_i), the volume of configuration bit-file packet (V_{cfp}) determined for the given reconfigurable PLD (e.g., partially reconfigurable FPGA), and the aforementioned configuration bus/port bandwidth (BW_{conf}) determined for this reconfigurable logic device:

$$T_{RPi\_conf} = \frac{m_i * V_{cfp}}{BW_{conf}} = \frac{m_i * V_{cfp}}{N_{cbpt} * F_{conf\_bus}} \tag{7.3}$$

If switching to the mode M_i is associated to reconfiguration of the RP RP_i, then switching to mode time $T_{Mi\_sw} = T_{RPi\_conf}$.

Obviously, the minimum mode-switching time for the particular reconfigurable logic device could be reached in the case when RP consists of one configuration frame. However, in most practical cases, such *elementary* RP is too small to accommodate a functional component. In most of cases, functional components will require an area consisting of several base regions and, therefore, a few configuration frames to program it into the reconfigurable PLD.

As an example, let us consider RP that consist of 10 configuration frames in Xilinx Virtex-6 FPGA. For this family of FPGA devices, the size of configuration frame bit-file is equal to 81 words × 32 bit/word = 2592 bits [2].

The SelecMAP-32 configuration port allows transmission of 32 bits per one transaction in frequency up to 100 MHz. Thus, the SelectMAP-32 bandwidth is equal to 32 bits × 100 MHz = 3,200,000,000 bit/s. Hereby, the reconfiguration time for a given RP in the Xilinx Virtex-6 FPGA will be T_{RPi_conf} = (10 frames × 2592 bits/frame)/3,200,000,000 bits/s = 6 microseconds.

In the case of 100 configuration frames included in the RP, the associated mode-switching time will be 64.8 microseconds accordingly. Thus, utilization of parallel configuration port and bus for Xilinx Virtex-6 FPGA device allows keeping the mode-switching time in the range of microseconds for RPs consisting from 2 to approximately 1234 configuration frames.

This can be compared to complete configuration time for Xilinx Virtex-6 FPGA devices. Indeed, the smallest Virtex-6 FPGA—XC6VLX75T—requires 26,239,328 bits for complete configuration [9]. Taking the same SelectMAP-32 port characteristics, the complete device configuration time will be equal to 26,239,328 bits/3,200,000,000 bits/s = 8.2 ms.

The largest FPGA in this family—XC6VLX760—[9] requires 184,823,072 bits for complete configuration and thus need 184,823,072 bits/3,200,000,000 bits/s = 57.76 ms.

In all cases, the complete reconfiguration of the entire FPGA device is in range of tens of milliseconds, which may be from two to three orders of magnitude higher than mode-switching time using RP reconfiguration. On the other hand, the serial configuration port is obviously much slower. However, the utilization of serial port and bus for the FCR configuration allows minimization of I/O pins used for configuration purposes. For example, initial complete configuration of the aforementioned Xilinx Virtex-6 FPGA, XC6VLX760, via serial configuration port at maximum configuration clock frequency equal to 100 MHz (excluding -1L Speed Grade [10]) will require 184,823,072 bits/100,000,000 bit/s = 1.848 s but need only four pins to be used for configuration in the case of JTAG or SPI port utilization.

Thus, complete and partial configuration and reconfiguration time may vary in big range from units of microseconds to units of seconds. This shortest reconfiguration time can be reached using parallel synchronous bus and port for configuration of small RPs in on-chip FCR and the longest configuration time for complete FCR configuration using serial synchronous port and bus. Hence, variation of configuration time can be up to six orders of magnitude that allows selection of proper configuration port and bus, RP size, configuration frequency according to the required mode-switching time, and area and power constraints.

7.7 Self-Reconfiguration of the On-Chip FCR and On-Chip Configuration Port

In the previous section, the initial complete configuration of the on-chip FCR and partial reconfiguration processes performed via the configuration port and bus were considered. However, there could be certain reasons for the partial reconfiguration of the FCR in run-time according to internal events. The main types of events are as follows:

1. Mode-switching request(s) caused by workload needs (already running tasks)
2. On-chip hardware faults caused by radiation, aging, and hidden manufacturing defects
3. Limited resources in the FCR causing temporal partitioning of available resources

In other words, there could be situations when the architecture configured in the on-chip FCR may need to modify itself according to certain on-chip events. This process can be considered as *self-reconfiguration of the on-chip system architecture.*

Obviously, initial complete configuration of the on-chip FCR is a must after power-up period of time. This complete configuration sets all interfaces and static logic configuration required by the application. In addition to that, initial system-on-programmable chip (SoPC) architecture also should be configured in respect to the initial mode of operation. Only after that, the reconfiguration for further mode(s) could be performed by partial reconfiguration.

The partial reconfiguration is usually initiated by the external configuration controller-loader (as discussed in Section 7.6). On the other hand, there should not be limitation to have the access to the on-chip configuration memory from the on-chip FCR. In this case, the internal configuration port must be included on the on-chip level of FCR. Thus, when reconfiguration of RPs is needed, this internal configuration port can be used avoiding external controller-loader and external configuration bus. Hence, this alternative way of partial reconfiguration of certain PRRs of the FCR may be performed faster, thus increasing SoPC flexibility as well.

As any partial reconfiguration process, self-reconfiguration of the on-chip FCR will require custom *internal controller-loader* (ICL) performing all write/read-back operations with partial configuration bit-files. This custom ICL must be configured in the static part of the SoPC during initial complete configuration of the on-chip FCR. In the case when on-chip configuration controller-loader should perform run-time partial reconfiguration of the RP(s), the on-chip configuration memory must provide access to any addressable configuration frames, that is, reconfigurable PLD should have an internal configuration port similar to external configuration port connected via configuration bus to the main memory storing all configuration bit-files for the application(s). The organization of such dual-ported access to the on-chip configuration memory is depicted in Figure 7.10.

In this approach, the reconfiguration mechanism consists of two parts:

1. External configuration mechanism (shown in the upper part of Figure 7.10)
2. Internal reconfiguration mechanism (shown in the lower part of Figure 7.10)

The well-known example of the internal configuration ports is the internal configuration access port (ICAP) in some of Xilinx FPGA devices [1]. This port is the internal port similar to Xilinx SelectMAP parallel configuration port discussed earlier. Thus, all functions of the SelectMAP-8/16/32 are available for the ICAP. Also, the organization of partial reconfiguration process is the same as the reconfiguration of RPs via SelectMAP port.

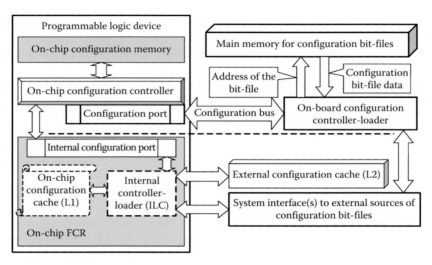

FIGURE 7.10
Dual-ported access to the on-chip configuration memory and configuration cache.

7.8 Organization of the On-Chip Configuration Cache Memory

One of the main reasons for the utilization of internal configuration port is the acceleration of partial reconfiguration and therefore faster switching to the new functions requested by application. This is possible to do because the on-chip data-transfer could be performed faster than at the onboard level. This is because the internal configuration controller can be designed with higher performance. Therefore, the higher bandwidth for internal configuration bus can be provided in comparison with the external configuration bus because it does not depend on external IOEs. Thus, the N_{cbpt}—the number of configuration bits per transaction—can be selected flexibly and optimized for the mode-switching time required by the application. The same is true for the F_{conf_bus}—on-chip configuration bus frequency that can be higher than the external configuration bus frequency. However, the reconfiguration period depends also on the bandwidth of the sources of configuration bit-files and the performance of control units involved in this reconfiguration process. To satisfy timing and power constraints, it would be better to have the source of configuration bit-files on-chip as well. Unfortunately, the volume of the on-chip memory (e.g., SRAM blocks in FPGA devices) is quite limited. Hence, it is not feasible to keep all configuration bit-files in the on-chip data memory units. The only way is to use these memory units as temporal storage for those configuration bit-files that are expected to be requested soon and/or often. In other words, the on-chip memory units keeping recently used configuration bit-files can be considered as on-chip configuration cache memory. This type of on-chip memory is exploiting the concept of temporal and spatial locality in a similar way as in conventional computers equipped with instruction cache. The effectiveness of utilization of cache memory in conventional (instruction-set-based) computers is based on concepts of temporal and spatial localities [11]. The *concept of temporal locality* assumes that probability of requesting again the functionalities (parts of algorithm) that have been required by application recently is very high. This is true for applications where the number of macrooperations can repeatedly swap each other depending on the mode of operation. The *concept of spatial locality* assumes that probability of requesting functionalities that are following the functions recently active in application also is very high.

Thus, the on-chip configuration cache can store the recently requested configuration bit-files as well as the configuration files associated with the functions referenced as consequent to active functions.

The utilization of concepts of temporal and spatial localities in the example of a video-processing application is illustrated in Figure 7.11. The example presented in this figure shows that after the initiation of the application segment, macrooperation MO_1 should be performed first. Thus, the associated configuration bit-file performing this macrooperation should be loaded to the dedicated RP of the configuration memory. Execution of this macrooperation will take certain time (e.g., 16.6 ms/video-frame according to 60 frames/second standard). During this time, two partial bit-files associated with MO_2 and MO_4 could be loaded in the on-chip cache as the only functionalities

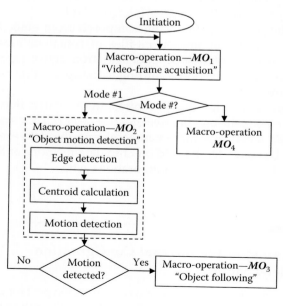

FIGURE 7.11
Temporal and spatial locality in the segment of video-stream processing.

that could be requested after buffering and color-decoding the video-frame. In the case when *Mode #1* is selected, the macrofunction MO_2 is activated by rapidly loading the associated bit-file to the addressed RP in the configuration memory. This RP configuration may be done even during the period of video-frame buffering to minimize the mode-switching time. Each macrooperation can contain the set of smaller macrooperations. As the example of MO_2 shown in Figure 7.11, it consists of three macrooperations: (1) edge detection, (2) centroid calculation, and (3) motion detection of the object.

Thus, the macrooperation can be divided into certain parts addressed to different RPs in the configuration memory. When *Mode #1* is selected, the probability of requesting MO_4 becomes very low. However, macrooperation MO_3 could be requested in the case when the motion of the object is detected. Thus, configuration bit-file associated with MO_3 should be loaded in the configuration cache memory instead of the configuration bit-file of the circuit performing MO_4.

It is necessary to mention that MO_2 being loaded in configuration memory still should be stored in the cache because in the active mode it can be requested soon again even when MO_3 is initiated for operation. This reflects the concept of *temporal locality* of the MO_2 that is in cycle in *Mode #1* and thus can be referenced again and again during this mode.

At the same time, the probability of referencing the MO_3 is very high when MO_2 is active because of logic dependency of MO_3 from MO_2. Indeed, the motion of the object can be detected in any next frame after the initialization of the MO_2. Thus, the aforementioned dependency represents the *spatial locality* (locality in "space") between MO_3 from MO_2.

Therefore, the on-chip cache being composed from embedded memory blocks can keep all configuration bit-files of functionalities that can be requested soon because of their dependencies from active segments of application or due to cyclic activation according to the application algorithm and/ or data structure.

It is important to mention that in contrast to the instruction cache in conventional processors, the on-chip *configuration cache can be composed inside the FCR dynamically* when this cache can accelerate the data-execution. Then, this cache can be converted back to on-chip data memory or buffer(s) when there is no need for the on-chip configuration cache.

7.9 Organization of the Internal Configuration Controller-Loader

Obviously, the utilization of the internal configuration port requires appropriate control and synchronization circuits for the configuration bit-file loading process. Thus, the ICL similar to the onboard configuration controller-loader is needed. This ICL can be configured in the FCR when needed and reconfigured to some other circuits when there is no further necessity in self-reconfiguration. Therefore, ILC and on-chip configuration cache are marked by dotted lines in Figure 7.10 due to their dynamic nature in the FCR. The ICL can be implemented in the FCR as a soft-core processor or as dedicated finite state machine (FSM)-based circuit (custom ICL). The most known implementation of the ICL in Xilinx FPGAs is based on MacroBlaze soft-core processor (e.g., [12]). Implementation of the ICL in soft-core processor simplifies the ICL design process and programming of all procedures associated with partial reconfiguration of the FCR over internal configuration port (e.g., ICAP) [13]. The drawback of this approach is reduction in mode-switching time due to relatively slow control and synchronization processes guided by instruction-set-based processor in comparison with dedicated control and synchronization circuits. The alternative implementation of the ICL could be the FSM-based custom dedicated circuit. This approach enables minimization of timing overhead caused by ICL and may require less hardware resources than in the case of soft-core implementation. The drawback of this approach is longer design time for the application-specific FSM-based ICL.

In general case, the ICL can directly access the on-chip configuration memory. However, in most practical cases, ICL, as well as the external configuration controller-loader, deals with the on-chip configuration memory via the on-chip configuration controller, a dedicated hard-core circuit, as shown in Figure 7.10. Actually, the ICL interacts with the on-chip configuration controller over the internal configuration port (e.g., ICAP) when the external configuration controller-loader interacts with the on-chip configuration controller via the external configuration port (e.g., SelectMAP). Thus, the internal configuration port often is just the internal (on-chip) "copy" of the external configuration port. As was mentioned earlier, the ICAP is just internal equivalent of the SelectMAP configuration port in several Xilinx families of FPGA devices.

From this point of view, the ICL is the on-chip functional equivalent of the external configuration controller-loader. Generally speaking, both devices should have similar functionality. The major functions of controller-loaders are the following:

1. Detection of the upcoming (requested or predicted) mode of operation
2. Selection of configuration bit-files of VHCs associated with detected mode of operation
3. Finding the location of selected bit-files in the main configuration memory
4. Determination of RPs in the on-chip configuration memory where bit-files of selected VHCs should be loaded
5. Scheduling the configuration bit-file loading procedures including

    a. Synchronization with ongoing data-execution process

    b. Determination of the safe moment for the start-up of the loading process of each of configuration bit-file to the respective RP

    c. Initiation and termination of the bit-file loading process from the external memory of configuration bit-files to the on-chip configuration controller

6. Interaction with on-chip configuration controller according to the selected communication protocol (e.g., JTAG, SPI)

7. Read-back configuration bit-file(s) and comparison with the original bit-files to detect errors caused by incorrect loading process or transient faults (e.g., single-event effects [SEEs] caused by radiation)

8. Initiation and termination of data-execution processes in the on-chip FCR and coordination of these processes with the existing embedded control circuits

All these functions can vary depending on the actual application specifics, the variants of RCS architecture, and the implementation of the configuration controller-loader. A more detailed discussion on these implementations and associated functions will be discussed in the next chapter in accordance with the variants of organization of the system level of reconfiguration control.

However, there is a conceptual difference between the internal and external configuration controller-loader. Since the initial complete configuration must be done in any way, the external controller-loader should be present in the system architecture in all cases. In contrast to that, ICL may not be included to the system architecture or may be configured in FCR for limited periods of time. The external configuration controller-loader can be implemented in different forms. For FPGA devices based on SRAM-configuration memory, it could be implemented as the external (to the FPGA device) unit based on CPLD, small FPGA, or the microprocessor. In some cases, for complete configuration of the SRAM-based FPGA devices, the dedicated controller-loader is embedded to the flash-memory based on the main configuration memory. In this case, one device combines both flash memory module containing configuration bit-file(s) and controller-loader that provides in-system programmability for FPGA devices. Some FPGAs can contain the embedded hard-core of configuration controller-loader and thus perform self-configuration by loading the initial complete configuration bit-file at power-up moment of time. These solutions providing in-system programmability can significantly simplify the system-level design and accelerate time to market. However, this solution may reduce flexibility of system adaptation excluding the ability for dynamic partial reconfiguration of the on-chip resources. The class of dynamically run-time reconfigurable RCS requires more complicated controller-loaders performing most of the earlier listed functions. These and other aspects of the system-level reconfigurability of the FCR will be considered in detail in the next chapter.

___

7.10 Summary

In this chapter, the main aspects of on-chip FCR reconfiguration were considered. It was shown that mode-switching time equal to reconfiguration could be significantly reduced in the case of partial reconfiguration of the on-chip resources. Therefore, the focus was put on the organization of the partial reconfiguration of resources and associated management of partial reconfiguration process in the on-chip FCR. Thus, partitioning of the on-chip FCR was discussed as well as the associated partitioning of the configuration memory. In accordance with this, complete configuration and further partial reconfiguration processes have been considered in conjunction to the structure of complete and partial configuration bit-files.

The bandwidth of configuration port and bus was analyzed as well as the main types of these ports and buses. The organization of the on-chip cache was presented with consideration in cases when this configuration cache can be effective. Then, functions of the configuration controller-loader were listed and discussed, putting attention on the interaction with the off-chip configuration memory via the configuration bus and system-level configuration controller-loader. This aspect, however, will be discussed in detail in the next chapter.

___

Exercises and Problems

Exercises

1. What are the two main parameters influencing the reconfiguration time of the FCR deployed in the on-chip level of the RCS (inside the PLD)?
2. What is the difference between PRR and BCR in the FCR?
3. How is BCR organized: (a) what are the components composing BCR and (b) what is its topology of components?
4. What is the smallest directly addressable segment of the on-chip configuration memory associated with the smallest configurable region of hardware resources in the on-chip FCR?
5. Define the correspondence between the resources in BCR and the associated configuration frame.

6. What is the RP in the on-chip configuration memory and how does it correspond with configuration frames?
7. What is VHC?
8. Can different VHCs target the same RP (a) at the same time or (b) at different periods of time?
9. What is the main reason why configuration bit-file of the VHC designed for allocation in one RP cannot be readdressed (in general case) to the other RPs?
10. List all the stages needed for the initial complete FCR configuration in the generic PLD.
11. What is the difference between the reconfiguration process of the generic RP and the complete FCR configuration in the PLD?
12. Why is the initiation procedure in the case of partial reconfiguration of an RP different from the initiation process of the entire on-chip FCR deployed in target PLD?
13. List the reasons that make the partial reconfiguration process of the RP much faster than the complete FCR in the PLD.
14. What are reasons that can make partial reconfiguration over the internal (on-chip) configuration port more effective than the reconfiguration via external configuration port? Which internal events can initiate the internal SoPC reconfiguration?
15. What is the main role of the configuration cache in the on-chip level of RCS? Should the configuration cache be present in the SoPC at all the time of its operation or can it be configured in SoPC for certain periods when this cache is needed?
16. What is the role of the ICL and in which forms it can be implemented in the SoPC?

Problems

1. The FCR in the video-processing RCS is deployed in the single FPGA device Xilinx XC5VLX85. The volume of the configuration bit-file for this FPGA device is equal to 21,845,632 bits. Calculate the bandwidth in MB/s. for the configuration port and channel (configuration bus) if the reconfiguration from one mode of video-processing to another should not exceed the period of one video-frame equal to 33.3 ms (in the case of frame rate equal to 30 frames per second).

   Note: 1 MB/s = $1024 \times 1024 \times 8$ bits/s

2. The volume of the configuration bit-file for the Xilinx XC5VLX85 FPGA device is equal to 21,845,632 bits. Out of this volume, 8704 bits are the encoding commands for on-chip configuration controller and other circuits composing bit-stream overhead.

   Calculate the number of configuration frames in this device, if the configuration frame length (for Virtex-5 FPGAs) is 41 words (32 bits per word).

3. The FCR of the RSC is deployed in the Xilinx XC5VLX85 FPGA device.

The configuration port selected in this FPGA is SelecMAP-32.

The configuration bus consists of 32 data-lines working on a frequency of 66 MHz.

The configuration frame length (for all Virtex-5 FPGAs) is 41 words (32 bits per word). The mode-switching time cannot exceed 100 ms for any mode of application.

Determine the *number of configuration frames* in the largest RP in the FCR that can satisfy mode-switching constrain for this application.

Note: The mode-switching time is assumed equal to the RP reconfiguration time.

References

1. Xilinx Inc., *Partial Reconfiguration*, User guide, UG702 (v14.1), Xilinx Inc., San Jose, CA, April 24, 2012, Chapter 6, pp. 89–100.
2. Xilinx Inc., *Virtex-6 FPGA Configuration*, User guide, UG360 (v3.8), August 28, 2014, Chapter 6, Table 6-23, p. 110.
3. Microsemi, IEEE standard 1149.1 (JTAG) in the SX/RTSX/SX-A/eX/RT54SX-S families, Application Note AC160, Microsemi, San Jose, CA, May 2012.
4. Texas Instruments, *KeyStone Architecture Serial Peripheral Interface (SPI)*, User Guide, SPRUGP2A, Texas Instruments, Dallas, TX, March 2012.
5. Xilinx Inc., *Virtex-6 FPGA Configuration*, User guide, UG360 (v3.8), August 28, 2014, Chapter 3, pp. 63–68.
6. ALTERA Corp., *Configuring Altera FPGAs*, CF51001, December 15, 2014.
7. Xilinx Inc., *Virtex-6 FPGA Configuration*, User guide, UG360 (v3.8), Xilinx Inc., San Jose, CA, August 28, 2014, Chapter 2, pp. 31–42.
8. ALTERA Corp., *Stratix V Device Handbook, Volume 1: Device Interfaces and Integration*, SV-5V1, ALTERA Corp., San Jose, CA, June 12, 2015, Chapter 8, pp. 8.1–8.16.
9. Xilinx Inc., *Virtex-6 FPGA Configuration*, User guide, UG360 (v3.8), Xilinx Inc., San Jose, CA, August 28, 2014, Chapter 6, Bitstream overview, Table 6-5, pp. 87–88.
10. Xilinx Inc., *Virtex-6 FPGA: Data Sheet*, User guides, and JTAG ID updates, XCN11009 (v1.0), Xilinx Inc., San Jose, CA, January 24, 2011.
11. J. L. Hennesy and D. A. Patterson, *Computer Architecture: A Quantitative Approach*, 3rd ed., Morgan Kaufmann Publishers, San Francisco, CA, 2003, Chapter 5, pp. 390–434.
12. T. Vladimirova and X. Wu, A reconfigurable system-on-chip architecture for pico-satellite missions, in *Communication Process Architectures 2007*, IOS Press, Amsterdam, the Netherlands, 2007, pp. 493–499.
13. J. Ayer Jr., Dual use of ICAP with SEM controller, Application Note: Virtex-6 and Spartan-6 Devices, Xilinx Inc., San Jose, CA, XAPP517 (v1.0), December 2, 2011.

8

RCS Architecture Configuration and Runtime Reconfiguration

8.1 Introduction

Chapter 7 was dedicated to the organization of reconfiguration mechanisms at the on-chip level of the FCR. The direct interaction between on-chip and off-chip reconfiguration mechanisms has been considered, including external configuration memory and respective controllers.

This chapter will describe different off-chip configuration and reconfiguration mechanisms at the system level of the RCS. This level may include the following elements and units:

1. Main memory for configuration bit-files and onboard configuration controller-loader

2. System interface(s) to the external sources of configuration bit-files

3. Temporal off-chip storages of configuration bit-files that can be considered as external configuration caches (level 2 [L2])

The initial configuration of multiple programmable logic devices (PLDs) should be discussed because, in general, FCR at its system level may include more than one reconfigurable device. Then, partial reconfiguration schemes will be considered for cases of partially and completely reconfigurable PLDs. Moreover, different implementations of the onboard controllers-loaders of configuration bit-files will be observed on the basis of existing design solutions. The benefits and drawbacks of each configuration and runtime reconfiguration schemes will be analyzed.

8.2 Methods of Start-Up Configuration of the FCR at a System Level

As was mentioned in Chapter 5, the system-level FCR consists of multiple PLDs. As PLDs, the FPGAs, CPLDs, or CGRAs could be considered. Each of these devices may require an initial complete configuration at the start-up time. Further reconfiguration during the operation may also be needed in case of functionality change.

First, let us consider the initial configuration process at the system level of the FCR. In general, there are four possible cases of the initial configuration of PLDs:

1. The initial configuration is preprogrammed in PLDs. No start-up configuration is needed.
2. Each PLD needs an individual start-up configuration different from other PLDs in the RCS.
3. All PLDs composing the FCR need the same initial configuration.
4. RCS consists of different groups of PLDs. PLDs in each group require the same initial configuration. However, the initial configuration of one group is different from others.

As per the concept of on-chip memory configuration described in Chapter 7, the configuration process could be performed via a serial or parallel configuration port and bus. Thus, the initial configuration of PLDs at the system level of the FCR can be organized via serial or parallel configuration ports in one of the following ways:

1. *Serial configuration of a single PLD*. In this case, the only reconfigurable device includes the FCR at a system level and should be configured at the start-up time by loading the initial configuration bit-file.
2. *Serial daisy-chain configuration of multiple PLDs* composing the FCR at a system level. In this case, two or more PLDs are configured by individual configuration bit-files transmitted from the main configuration memory source to each PLD via a serial configuration bus organized as a chain from one PLD to another.
3. *Serial ganged configuration of multiple PLDs* composing the system-level FCR. This case assumes that all PLDs must have the same initial configuration. Thus, one main configuration memory source can be used to load the same complete configuration bit-file to all the PLDs simultaneously over the serial configuration bus.
4. *Parallel configuration of a single PLD*. This case is applicable for systems consisting of one RLD at a system level of the RCS architecture that requires fast initial configuration.

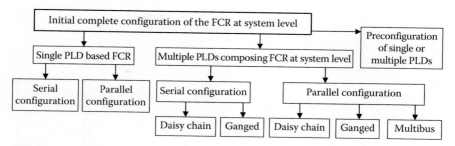

FIGURE 8.1
Classification of initial complete configuration of the FCR at a system level of the RCS.

5. *Parallel selected configuration of multiple PLDs* allows loading the complete configuration bit-file via the parallel configuration bus to the respective PLDs. If configuration is conducted sequentially one PLD after another, then scheme is considered as *parallel daisy chain*.

6. *Parallel ganged configuration of multiple PLDs* composing the system-level FCR. This case is similar to serial ganged configuration but using parallel configuration bus.

7. *Parallel multibus configuration of multiple PLDs.* In this scheme, multiple configuration buses are individually connected to each PLD from the controller-loader. This scheme provides the shortest time for the initial RCS configuration but requires maximum additional hardware resources at a system level of the RCS architecture.

All the aforementioned methods of initial configurations of PLDs in the RCS are presented in Figure 8.1.

The initial configuration of the FCR deployed in the *single PLD* was already discussed in Chapter 7. Thus, only the methods used for the complete configuration of multiple PLDs composing the system level of the FCR will be considered further.

8.3 Serial Daisy-Chain Configuration Scheme

This scheme assumes that each PLD (mostly a CPLD or FPGA device) allows passing the configuration data through the on-chip circuits from the configuration data-in port to the data-out port. In most cases, this scheme is used for the serial configuration of two or more FPGAs or CPLDs. Quite often, there could be a mixed chain of configurable devices. However, in such cases, timing and protocol constrains must be supported by all devices

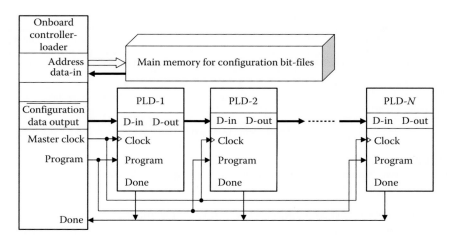

FIGURE 8.2
Daisy-chain configuration scheme for complete configuration of multiple PLDs.

in the chain. The general organization of the daisy-chain scheme for multiple PLDs is depicted in Figure 8.2.

Serial daisy-chain programming can be implemented in two modes: (1) passive or (2) active. In the *passive mode*, the external configuration controller-loader of the PLD must be used as it is shown in Figure 8.2. In the *active mode*, also called the *master mode*, the internal (on-chip) configuration controller of the master PLD plays the role of a complete configuration bit-file loader. In this case, the on-chip configuration controller directly interacts with the main (external) memory for configuration bit-files providing address of the configuration bit-file data, configuration clock, and necessary initiation and feedback signals. Therefore, the PLD can conduct a self-configuration as well as a daisy-chained configuration of *slave* PLDs connected to the *master* PLD. The self-configuration scheme of the master and slave PLDs connected in the daisy chain is shown in Figure 8.3.

In initial (power-up) moment of time, the "Program" signal (pin) should be asserted. This signal indicates the start-up moment of time for single- or multi-PLD configuration. Then, the master PLD starts the generation of the clock to the main memory containing configuration bit-files and all slave PLDs. Configuration data come from the main configuration memory to the master PLD first. When the on-chip configuration memory of the master PLD is filled, the on-chip configuration controller of this PLD switches the configuration data-stream to the D-out of the device and configuration of the next PLD starts. This process continues until the last PLD in the chain completes its configuration and the last "Done" signal confirms correctness of receiving the configuration bit-file. Therefore, it is possible to consider the serial daisy-chain configuration process as a bit-by-bit programming of the configuration memory of multiple PLDs as one large configuration memory

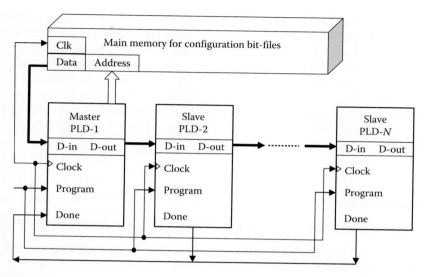

FIGURE 8.3
Self-configurable daisy chain of master PLD-1 and slave PLDs.

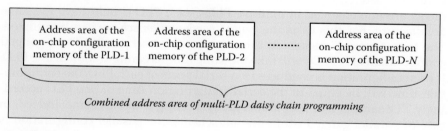

FIGURE 8.4
On-chip configuration memory address area for daisy-chain programming.

combined from the on-chip configuration memories of each PLD in the RCS, as shown in Figure 8.4.

This approach simplifies the design of the onboard and system levels because no additional configuration controller-loaders need to be added for individual configuration of each PLD in the system. Certainly, implementation of this method requires additional on-chip circuits performing (1) configuration control and (2) through-the-chip configuration data-transfer. Further, the software CAD support must provide generation of combined configuration bit-file for *N*-PLDs according to the order of PLDs in the daisy chain. In turn, this method enables the initial configuration of different types of PLDs connected in the chain (e.g., FPGAs and CPLDs).

Summarizing the aforementioned, the daisy-chain configuration scheme for multiple PLDs obviously is the most *economical* in terms of additional onboard hardware resources. However, this scheme is slowest in terms of the start-up time of the system due to the serial nature of the bit-by-bit configuration data-transfer to each of the PLDs connected sequentially as well. Thus, this scheme is efficient for implementation in the RCS where rapid start-up is not critical. Most of the FPGA/CPLD vendors provide daisy-chain configuration support in their products (e.g., Xilinx Inc. [1], Altera Corp. [2]).

8.4 Serial and Parallel Ganged Configuration of Multiple PLDs

In cases when the application requires a start-up time exceeding the initial configuration time provided by the daisy-chain scheme, other types of configuration schemes should be considered.

In many applications where large massive of data must be executed in a relatively short time, the array of identical processing elements working in parallel is usual architectural solution. In cases when these processing elements are implemented in multiple PLDs, each PLD should have the same architecture. For this case, the *ganged configuration scheme* can be utilized. The term "ganged" means, in this case, *multichannel parallel*. Thus, ganged configuration of N PLDs will take the same time as configuration of one PLD. If this configuration is conducted via serial ports of each PLD, the configuration time will be equal to the serial configuration time of one PLD accordingly. The same is true for parallel ganged configuration scheme. Indeed, in case of parallel ganged configuration scheme, the parallel configuration is performed via parallel ports of each PLD. In other words, ganged configuration of N PLDs accelerates the initial configuration process in N-times without much additional hardware resources. The passive serial/parallel ganged configuration scheme is depicted in Figure 8.5.

Figure 8.5 shows that the configuration data bus connects the controller-loader's configuration data-output port with the D-in ports of each PLD in parallel via the configuration data bus. If this bus is serial (1-bit data-transfer line), the ganged configuration is serial. Otherwise, when the configuration bus is word-wide, the ganged configuration scheme is parallel. Obviously, parallel ganged configuration scheme provides acceleration of the initial RCS configuration equal to the width of the configuration bus. For example, if the configuration bus consists of 32 data transmission lines, the RCS will be configured 32 times faster than in the case of a serial ganged configuration for the same configuration bus frequency. Thus, the total acceleration of the configuration process in comparison with the daisy-chain scheme is equal to $N * W_{conf\_bus}$, where N is the number of PLDs in the array and W_{conf_bus}

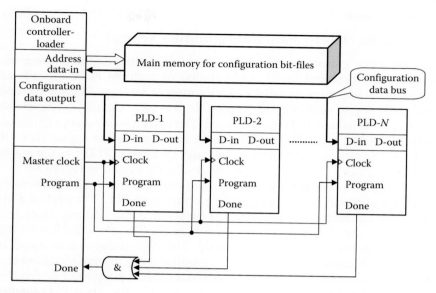

FIGURE 8.5
Serial/parallel ganged configuration scheme for multi-PLD RCS.

is the width of the configuration bus. For example, if there are 16 PLDs in an array and a 32-bit bus is used, the total acceleration of the configuration process in comparison with the serial daisy-chain configuration is equal to 16 * 32 = 512 times.

It is also necessary to mention that there is no additional hardware resource overhead required for implementation of this scheme except the feedback circuit (e.g., "Done"). In this scheme, all "Done" signals should be collected in one AND gate with N inputs and one output connected to the "Done" input of the controller-loader. Thus, configuration process is completed when all PLDs have reported their successful configuration. Then, the RCS can proceed to the user mode and start running the application. If the error is detected in any PLD in the configuration process, the configuration process of all PLDs should be repeated entirely. This aspect can be considered as the drawback of this scheme of the initial RCS configuration. However, the same problem exists for the daisy-chain scheme too. Thus, reliability and noise immunity for the configuration bus should be considered as one of the most important aspects in onboard and system designs.

Summarizing the considered ganged scheme for the initial complete configuration of multi-PLD RCS, it is possible to see that a significant acceleration of the configuration process could be reached with a quite limited additional hardware resources. However, this scheme is applicable only for arrays of identical processing elements when the number of PLDs has the same architecture with their on-chip FCRs.

8.5 Parallel Daisy-Chain Configuration Scheme

In cases when ganged configuration is not applicable due to the specifics of the application and serial configuration of multiple PLDs exceeds the required start-up time, parallel programming of multiple PLDs is the only way to accelerate system configuration. In this case, parallel configuration bus and ports in each PLD must be used. The daisy-chain concept still could be implemented if all PLDs in the chain can provide the appropriate I/O support. The self-configurable parallel daisy chain with PLD-1 as the master device and PLD-2 … PLD-*N* as slave devices is illustrated in Figure 8.6.

This configuration scheme allows rapid configuration of *N* PLDs using parallel configuration bus and parallel ports in each PLD. The control of the process is conducted by the first PLD by sending the address, clock, and associated control signals to the main memory for configuration bit-files (usually nonvolatile memory like flash). The configuration data word by word are transmitted to all PLDs via a parallel configuration bus. However, only one PLD receives the configuration data that is supported by the chip select (CS) signal. First, the master PLD-1 receives its file. Then, PLD-1 generates the chip select output (CS-O) signal. This signal comes to the chip select input (CS-I) of the next-in-chain PLD. This allows PLD-2 to receive the configuration bit-file addressed to this device. When configuration of PLD-2 is completed, it generates its own CS-O signal to the next PLD and this process continues until all PLDs are configured. The integrated configuration bit-file should be compiled as a composition of appropriate configuration bit-files

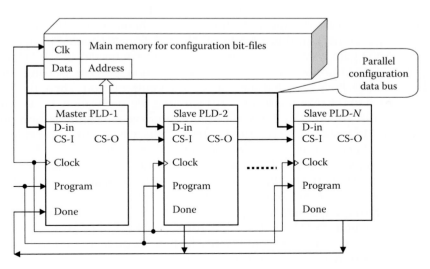

FIGURE 8.6
Self-configurable parallel daisy chain of *N* PLDs with PLD-1 as the master.

in the order of PLDs in the daisy chain. The acceleration of configuration process in comparison with a serial daisy-chain scheme is equal to the width of the parallel configuration bus—W_{conf_bus}. For instance, if $W_{conf\_bus} = 32$, the configuration will run 32 times faster than the serial daisy-chain scheme. It is also necessary to mention that this scheme requires only a wider bus and can be performed by one of the PLDs in case the appropriate circuits are embedded in its microarchitecture. Fortunately, most FPGA vendors provide the previously mentioned features in their FPGA devices (e.g., Xilinx Inc. [3], Altera Corp. [4]).

8.6 Parallel Passive Configuration Scheme

In cases when the PLD does not support daisy-chain circuits as well as self-configuration features, the parallel configuration scheme still is possible to be used. However, the external (onboard) configuration controller-loader should be added for this configuration scheme. All PLDs must be in passive configuration mode as slave devices. The configuration controller-loader is working as a master device retrieving configuration bit-files from the main configuration memory and sending them via a parallel configuration bus to the appropriate PLD. Because the configuration bus is common bus for all PLDs, selection of the target PLD is done by assertion respective CS signal. Thus, the configuration controller-loader can address the configuration bit-file to the associated PLDs in any order (not necessarily chain). Therefore, the additional configuration controller-loader allows the same acceleration rate of the configuration process as in the parallel daisy-chain configuration scheme but for almost any PLDs and in any order of configuration sequence.

This configuration scheme is depicted in Figure 8.7.

8.7 Multibus Configuration Scheme for Multiple PLDs

In some cases, the application may require minimization of the start-up time. In such cases, the maximum parallelism in the configuration scheme should be utilized at the highest possible bus and port frequency. This assumes that all PLDs in the RCS should be configured simultaneously using the widest possible parallel configuration bus and highest bus frequency. The organization of this scheme of system configuration is shown in Figure 8.8.

This scheme allows simultaneous configuration up to N PLDs in any order. The onboard controller-loader is a master device generating a master clock

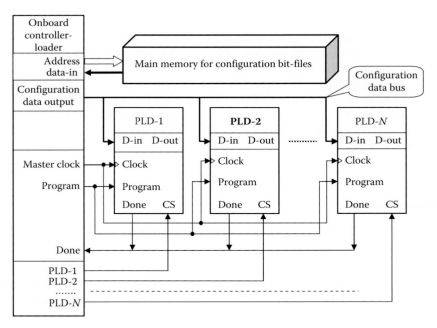

FIGURE 8.7
Parallel passive configuration scheme of *N* PLDs with master controller-loader.

and all necessary control signals to PLDs at the system level of the RCS. Also, controller-loader generates the addresses of initial configuration bit-files for the main configuration memory. In this scheme, the main configuration memory usually consists of *N* nonvolatile configuration memory units (e.g., flash memory devices). Each of the memory units contains the complete configuration bit-file for the respective PLD. All configuration memory units are connected to the common address bus. The data outputs of the previously mentioned memory units are connected to the configuration ports of the respective PLDs via individual parallel configuration buses. Therefore, this scheme provides the highest possible bandwidth for configuration information and, thus, the shortest start-up time for the RCS.

From a quantitative point of view, if RCS consists of *N identical PLDs* with the same configuration ports, configuration bus width (W_{conf_bus}), and configuration bus frequency (f_{conf}), the complete configuration time (T_{conf}) of all PLDs is equal to the configuration time of *one* PLD in this system:

$$T_{conf} = \frac{V_{pld}}{W_{conf_{bus}} \times f_{conf}} \qquad (8.1)$$

where V_{pld} is the volume of configuration bit-file for every PLD in the set.

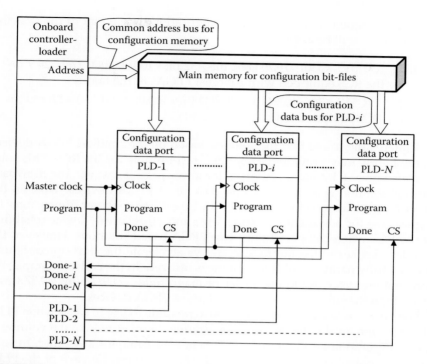

FIGURE 8.8
Multibus configuration scheme with master controller-loader and N slave PLDs.

As an example, let us consider the RCS based on Xilinx Virtex-5 FPGA family discussed earlier.

If the application requires for implementation (1) 100,000 LUT-based logic slices, (2) 360 DSP slices, and (3) block RAM volume equal to 20 MB, it can be implemented on the following:

1. Two largest XC5VLX330 FPGA devices providing $2 \times 51,840 = 103,680$ LUT slices, $2 \times 192 = 384$ DSP slices, and $2 \times 10,368$ kB $= 20,736$ kB $= 20.25$ MB or

2. Twenty-one smallest XC5VLX30 devices providing $21 \times 4,800 = 105,600$ LUT slices, $21 \times 32 = 672$ DSP slices, and $21 \times 1,152$ kB $= 24,192$ kB $= 23.625$ MB

The volume of the configuration bit-file for the largest XC5VLX330 is equal to 79,704,832 bits when the volume of the configuration bit-file for the smallest XC5VLX30 is equal to 8,374,016 bits.

If a fully parallel configuration scheme is selected for this RCS using the fastest SelectMAP-32 (32-bit-wide) configuration port and bus working at a

100 MHz frequency, the complete configuration time equal to the system's start-up time will be as follows:

1. T_{conf} (XC5VLX330) = 79,704,832 bits/(32 bit × 100,000,000 1/s) = 24.9 ms for the RCS equipped with 2 largest FPGAs in the family
2. T_{conf} (XC5VLX30) = 8,374,016 bits/(32 bit × 100,000,000 1/s) = 2.6 ms for the RCS equipped with 21 smallest FPGAs

It is interesting to see that the solution based on the smallest FPGA devices in the family may provide the shortest start-up time for the RCS. This solution also gives a bit more resources for lower price. However, the drawback for this solution is a larger area needed for 21 FPGA devices. This, on the other hand, causes (1) higher cost of the board due to larger dimensions; (2) higher cost for assembling, testing, etc.; and (3) relatively lower reliability of the RCS when compared with two FPGA-based solutions. However, the smallest devices in the family can always accelerate the start-up configuration if a fully parallel multibus scheme is utilized. In the earlier example, the acceleration was close to one order of magnitude—24.9/2.6 ms = 9.57 times faster than the solution based on two largest FPGA devices.

The same configuration scheme is also possible to use in cases when PLDs at the system level of the RCS *are not identical*. In other words, the volume of the configuration bit-file for one PLD is not the same for any other. In such case, the controller-loader must provide the CS signals for each of the PLDs in the set. Thus, the previously mentioned CS signals are set at the beginning of the configuration process but are de-asserted individually according to the "Done" feedback signal received from the respective PLD. This scheme is shown in Figure 8.8.

The period of the RCS complete configuration will be equal to the maximum configuration time of the largest PLD in the set. In cases where the width of the configuration bus (W_{conf_bus}) and configuration bus frequency (f_{conf}) is the same for all PLDs, the configuration time for such RCS is

$$T_{conf} = \frac{\max\{V_{pld_i}\}}{W_{conf\_bus} \times f_{conf}} \tag{8.2}$$

where V_{pld_i} is the volume of configuration bit-file for the PLD_i, $i = 1, 2,\ldots, N$.

It is necessary to mention that the design of the controller-loader used in this scheme can be more complex than in the case of identical PLDs. It may require additional small FPGAs because resources deployed in CPLD devices often are not enough to provide adequate hardware support for the earlier scheme when the number of PLDs at the system level of the RCS is relatively big.

In all cases, the multibus configuration scheme can provide the shortest configuration time for RCS with multiple PLDs, requiring, however, more hardware resources for its implementation.

8.8 Preconfiguration of Single or Multiple PLDs

In almost every RCS architecture, there is a necessity to have one or more PLDs (e.g., FPGAs or CPLDs) already preconfigured at the start-up time. These preconfigured PLDs have nonvolatile configuration memory instead of SRAM-based memory in its architecture. The nonvolatile memory used in these PLDs could utilize different technologies such as EPROM, EEPROM, NOR, or NAND-flash memories or even fuse/anti-fuse-based memory elements.

In all cases, the configuration of the previously mentioned PLDs is stored in the devices when power is shut down, and thus, there is no need to restore it at start-up. Hence, their configuration can be considered as immediate after power-up.

There are many reasons why RCS needs such preconfigured PLDs in its architecture. For example, the configuration controller-loaders based on PLDs and described in previous sections should be ready to perform their functions right after powering up the RCS. The I/O device drivers and controllers are another example of such PLDs. In other words, all devices based on PLDs that need to be ready to work right after powering the RCS must be preconfigured before the initiation of RCS. The reconfiguration of these PLDs may still be needed to update their functionality. For this purpose, most PLDs are based on EEPROM or flash memory devices with JTAG access for serial reconfiguration (e.g., Xilinx CPLDs [5], Xilinx FPGAs [6], Lattice FPGAs [7]).

8.9 Organization of Runtime Reconfiguration of the FCR at a System Level

As was discussed in Chapter 7, in many multimodal applications, further reconfiguration of the FCR in each PLD may be needed in runtime. For this purpose, the FCR at the system level of the RCS may consist of multiple PLDs, each of which may have a partially reconfigurable architecture. It is necessary to mention again that from a system point of view, runtime reconfiguration could be organized even if PLDs have a statically configurable architecture (nonpartially reconfigurable). However, if the application is deployed on multiple PLDs and some PLDs are in the process of reconfiguration while the rest of PLDs still are processing the current workload, such organization of the reconfiguration process can be considered as runtime reconfigurable FCR at the system level of the RCS. In other words, the system level of partial reconfiguration just extends the granularity of reconfigurable units to PLD size from the smallest on-chip level of the base configurable region (BCR) inside the partially reconfigurable PLD.

There could be a large variety of runtime reconfiguration schemes and associated mechanisms performing runtime partial reconfiguration at every level of the FCR from the on-chip to the system level. However, most of them could be divided into two large classes:

1. Runtime complete PLD reconfiguration schemes when one or more PLDs are reconfigured while other PLDs are in the data-execution process
2. Dynamic partial reconfiguration of reconfigurable regions of the on-chip resources

The complete runtime reconfiguration of the PLD in-system level of the RCS can be organized as self-reconfiguration (e.g., multiboot scheme in some Xilinx Spartan-6 FPGA family [8]) or as externally controlled complete device reconfiguration. This type of reconfiguration is used for multimodal or multifunctional statically reconfigurable RCS such as discussed in Section 6.3.

The dynamic partial reconfiguration of the on-chip resources can also be organized as a self-reconfiguration process under the control of an on-chip reconfiguration controller or by the external control of a dynamic reconfiguration process using an external controller-loader. The reconfiguration control can be centralized (performed by one central controller) or distributed between PLDs.

All the earlier cases are applicable for single PLD–based RCS as well as the multi-PLD RCS.

The classification tree of the previously mentioned possible variants of runtime reconfiguration of the FCR at the system level of the RCS is depicted in Figure 8.9.

The previously listed variants of runtime reconfiguration schemes will be discussed in the following sections in detail.

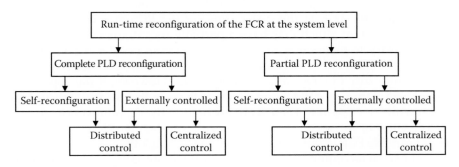

FIGURE 8.9
General classification of runtime reconfiguration schemes at the system level of the RCS.

8.10 Multiboot Runtime Self-Reconfiguration of Single and Multiple PLDs

This scheme allows complete reconfiguration of a PLD to switch its functionality from one to another by self-loading the requested bit-file from the external configuration memory. This scheme could be used for multimodal statically reconfigurable RCS such as discussed in Section 6.3. For this scheme, the target PLD must have embedded dedicated hardware circuits in its microarchitecture to support this type of reconfiguration. The on-chip hardware support should have the mode-selection register to keep the requested mode number before resetting the current configuration of the PLD. Thus, the content of this register is associated with the ID-code of the requested bit-file. The ID-code of the bit-file presents the upper (most significant) bits of the address area of external configuration memory. The general organization of the aforementioned multiboot scheme for a single-PLD self-reconfiguration is shown in Figure 8.10.

In the case of RCS with multiple PLDs at the system level of the FCR, the multiboot reconfiguration allows easy implementation of multimodal operation of the RCS by runtime reconfiguration of one or few PLDs while the rest of PLDs continue their data-execution processes. Let us assume that RCS is employing *N*—PLDs at the system level—and mode switching requires changing functionality of at least one PLD in the system. If the

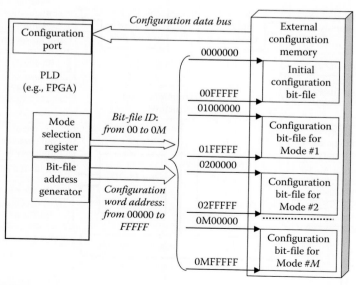

FIGURE 8.10
Multiboot self-reconfiguration scheme for single PLD.

PLD$_i$, $i = 1, 2, \ldots, N$, can work in one of the M_i modes, the total number of possible modes in such RCS is equal to $M = \Pi_1^N M_i$.

In this case, the total number of configuration bit-files to be generated and stored in the external configuration memory module(s) is equal to $\Sigma_1^N M_i$, $i = 1, 2, \ldots, N$, which is much lesser than the possible modes of operation available in such RCS organization.

For example, if the RCS contains four PLDs and each PLD may be configured in one of the four configurations, the total number of modes accommodated in this RCS is equal to $4^4 = 256$ modes. However, the number of configuration bit-files to be stored in the main configuration memory is equal to $4 \times 4 = 16$. If the number of configurations for each PLD is increased to 8 and, thus, the total number of configuration bit-files increases up to $4 \times 8 = 32$, the total number of possible modes will increase to $8^4 = 4096$. Hence, doubling the number of configuration bit-files allows increasing variations of operation modes up to 16 times. In other words, doubling the volume of the external configuration memory allows increasing the flexibility of this RCS 16 times using the same PLDs and with runtime mode switching ability. The example of the previously mentioned configuration scheme is depicted in Figure 8.11 representing the RCS with four PLDs, each of which is connected to its own external configuration memory module. Coordination of mode functionality changes in each PLD should be conducted by *reconfiguration management system (RMS)*. This system can be

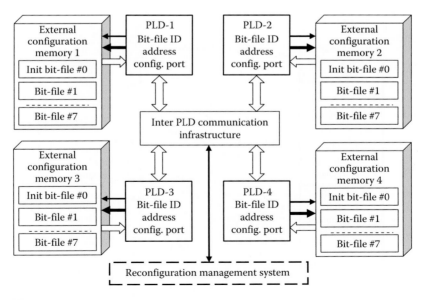

FIGURE 8.11
Multiboot runtime self-reconfiguration scheme for multi-PLD RCS.

implemented as (1) a distributed system using inter-PLD communication or (2) centralized system using master reconfiguration controller. This aspect will be considered in Chapter 10.

In addition to the aforementioned benefits, the multiple configuration buses in this scheme can provide parallel start-up configuration of all PLDs. This is similar to the multibus configuration scheme discussed in Section 8.7 and presented in Figure 8.8.

At the same time, a multiboot scheme does not require an onboard controller-loader since the on-chip hardware circuits can provide all necessary addressing and bit-file loading operations.

However, practical implementation of the previously mentioned scheme is possible only in cases when the microarchitecture of the PLD provides such hardware support. As an example of the previously mentioned PLDs, Xilinx Virtex-6 FPGA devices providing a *fallback multiboot* reconfiguration can be considered [9].

8.11 Multiboot Runtime Reconfiguration with Distributed External Control

In cases where the multiboot self-reconfiguration hardware support is not available in a given PLD, it is still possible to implement the previously mentioned scheme of runtime reconfiguration using centralized or distributed external control.

Distributed control scheme assumes just an additional controller-loader for each pair of PLDs and respected external memory module. This scheme is shown in Figure 8.12.

This scheme is similar to the previous scheme of multiboot self-reconfiguration for multi-PLD RCS. The difference is that hardware support for runtime reconfiguration of complete PLD(s) is provided by the external controllers instead of the embedded reconfiguration control circuits in the previous scheme. Thus, mode registers generating the configuration bit-file ID code and associated address registers as well as counters dealing with the respective external configuration memory modules are allocated in the external (to target PLD) controller-loader dedicated to the associated PLD. As in the previous multiboot scheme, coordination of the runtime reconfiguration process is a task of the RMS. This system can be centralized or distributed as well as reconfiguration mechanism.

The aforementioned scheme provides several benefits for the RCS employing multiple PLDs at the system level of the FCR. First, this scheme makes possible the utilization of relatively inexpensive FPGAs as PLDs in comparison with FPGA devices that contain the multiboot self-reconfiguration

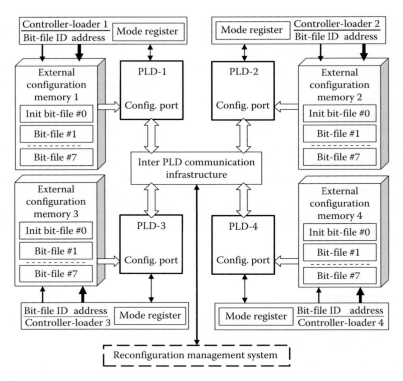

FIGURE 8.12
Multiboot runtime reconfiguration scheme with distributed external control.

hardware. The PLD cost could be lowered sometimes in order of magnitude by utilizing multiple but inexpensive reconfigurable devices instead of one large and expensive PLD.

The additional external controller-loaders based on inexpensive CPLDs (e.g., Xilinx CoolRunner [5] or similar) in most practical cases may increase the cost of the system in percentage in comparison with the cost of FPGA devices. For example, the unit cost of the Xilinx CoolRunner-II CPLD with 128 macrocells and 100 input/output pins in the 144QFP package does not exceed ~$10 in comparison with hundreds or even thousands of dollar cost of the associated FPGA devices.

On the other hand, the distributed reconfiguration control can provide parallel initial and runtime reconfiguration of multiple PLDs independently. This scheme, as any distributed control scheme, also increases reliability, flexibility, and testability of the reconfiguration system of the RCS.

One of the drawbacks of this scheme is the relatively larger printed circuit boards (PCBs) required to accommodate the additional controller-loaders. However, quite often, allocation of the controller-loaders on the opposite side of the PCB under the external configuration memory modules (flash memory units) allows keeping the PCB dimensions close to or the same as the dimensions of PCBs deploying PLDs with multiboot self-reconfiguration abilities.

Another limitation of both previously mentioned schemes is the inability to use bit-files stored in the external bit-file memory dedicated to PLD_i for any other PLD_j, $i \neq j$, $i, j = 1, 2, \ldots, N$.

Obviously, the previously mentioned schemes assume dedication of the i-set of bit-files to the PLD_i only. This assumption is based on the fact that PLD_i and PLD_j may be different types of PLDs or even produced by different vendors. In such cases, the aforementioned schemes can be considered as the most cost-effective or even the only possible solution when start-up and runtime reconfiguration periods should be minimized.

8.12 Multiboot Runtime Reconfiguration with Centralized External Control

However, in cases when all PLDs are identical and application does not require minimization of the start-up time or runtime reconfiguration periods, the centralized multiboot configuration scheme could be considered as the potential solution. This scheme is quite identical to the parallel passive configuration scheme described in Section 8.6 for the initial configuration of multi-PLD RCS and is depicted in Figure 8.7. Actually, this scheme can be easily modified to perform the runtime reconfiguration of any PLD in the system when other PLDs are continuing their data-processing. The modification requires deployment of mode registers associated with each PLD in this system and the mode-control bus interconnecting all PLDs with the master controller-loader.

The master controller-loader is the centralized configuration and reconfiguration control unit performing the following functions:

1. *Start-up configuration* of the initial mode of operation of all PLDs in the system. This start-up configuration is done by setting the start address of the configuration bit-file for each PLD, asserting the CS signal for this PLD, and conducting the bit-file loading process according to the generated master clock. Thus, the start-up time in this scheme is N-times longer than in schemes discussed in Section 8.10 and 8.11.

2. *Runtime reconfiguration* of one PLD (at a time) while other PLDs are performing the task execution (Figure 8.13). In this case, the master controller-loader sets the start address of the bit-file to be loaded in the selected PLD and asserts the CS signal for the selected PLD. The initiation of this procedure should be coordinated with (a) PLDs

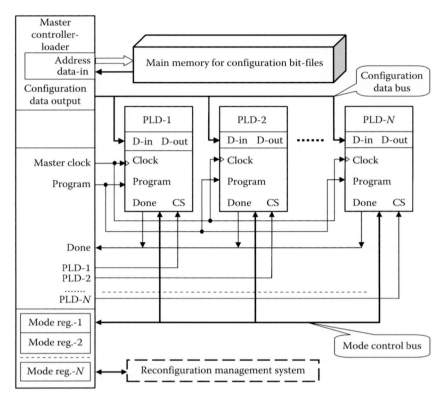

FIGURE 8.13
Multiboot runtime reconfiguration scheme with centralized external control.

continuing their data-processing and (b) the RMS (part of operating system of the RCS) to avoid possible conflict(s) in the data-execution process. This coordination can be done using the mode control bus allowing information exchange between the master controller-loader and the associated circuits in each PLD.

Since the configuration data bus and mode control bus are shared between PLDs, the communication and further reconfiguration of PLDs can be done only in a one-by-one basis. In turn, the runtime reconfiguration process could be done only by complete reconfiguration of one PLD when in previous multiboot schemes several PLDs can be reconfigured simultaneously. However, the advantage of centralized scheme of reconfiguration control is the ability to load any configuration bit-file to any PLD in the system. Therefore, this scheme, being slower in terms of start-up and runtime reconfiguration, is certainly more flexible in terms of utilization of configuration bit-files.

On the other hand, centralized control reduced reliability and survivability of the system. In this case, any hardware fault in the master controller-loader of the main memory for configuration bit-files causes catastrophic failure for the entire RCS.

It is necessary to mention that the master controller-loader can be implemented in (1) CPLD or FPGA with nonvolatile configuration memory and/or (2) RISC controller coupled with CPLD. The second solution allows easier interfacing of the master controller-loader with external networks to upload or update the set of configuration bit-files on the system. Another practically useful technique is doubling or even quadrupling the memory modules storing the configuration bit-files. This allows increasing the reliability of the system as well as reconfiguration bandwidth in cases when flash-based memory devices cannot provide enough bandwidth required for the configuration port(s) of PLDs. The example of such scheme is presented in Figure 8.14.

8.13 Partial Runtime Reconfiguration with Centralized External Control

The aforementioned multiboot reconfiguration schemes considered the entire PLD as the smallest granularity of the FCR for the runtime reconfiguration. This approach allows utilization of PLDs that do not have partial reconfiguration abilities. The potential drawback is that it may require an unacceptable period for mode switching in comparison with the approach based on the partial on-chip reconfiguration of FCR. This aspect was considered in detail in Section 7.6.

The runtime reconfiguration scheme with centralized control discussed earlier in Section 8.12 makes possible a significant reduction in the reconfiguration time if PLDs with partial reconfiguration abilities are used. The master controller-loader in this case performs the same function for the initial configuration of all PLDs as in the multiboot configuration scheme. However, for the runtime reconfiguration, the master controller-loader can reconfigure the entire PLD as well as its on-chip partially reconfigurable regions (PRRs) inside the selected PLD.

The general organization of this scheme is the same as shown in Figures 8.13 or 8.14. Obviously, the main configuration memory in this case should contain partial bit-files associated with mode changes instead of complete configuration bit-files as it was necessary in multiboot runtime reconfiguration schemes. Also, reconfiguration procedures performed by the master controller-loader should reflect partial configuration bit-file loading processes discussed in Section 7.5 and depicted in Figure 7.8. This scheme can

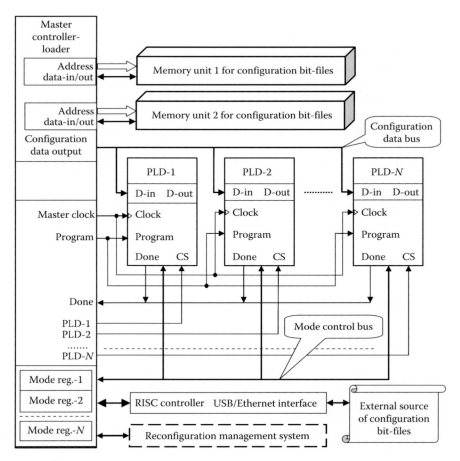

FIGURE 8.14
Multiboot runtime reconfiguration scheme with dual configuration memory units.

be considered as an *externally controlled partial reconfiguration scheme with centralized control.* The scheme provides high flexibility in RCS reconfiguration, being able to use any bit-file stored in the main configuration memory for any PLD in the system. At the same time, this scheme allows significant reduction in reconfiguration time for a PLD due to partial bit-file loading instead of complete PLD reconfiguration. In turn, it enables minimizing the mode switching time even when more than one PLD should be reconfigured for the upcoming mode. On the other hand, all potential problems associated with centralized reconfiguration control are the same for this reconfiguration scheme as for the multiboot scheme with centralized control discussed earlier. In addition to that, it is necessary to mention that PLDs allowing partial reconfiguration of their on-chip FCR are more expensive in comparison with PLDs designed only for complete reconfiguration.

In practical applications, the multiboot reconfiguration scheme with centralized control usually is used for relatively small and inexpensive FPGA devices. This allows minimization the RCS cost and reconfiguration time due to relatively small volume of configuration bit-files for these FPGAs. Alternatively, utilization of partially reconfigurable FPGA devices may provide shorter mode switching time with less FPGA devices on board. There is always some trade-off between the aforementioned possible runtime reconfiguration schemes.

8.14 Partial Runtime Reconfiguration with Distributed Control

Decentralized organization of runtime partial reconfiguration control can mitigate many drawbacks caused by the centralized scheme. First, this scheme of reconfiguration reduces mode switching time due to ability for simultaneous reconfiguration of multiple PLDs involved in mode-change process. Therefore, the mode switching time may not exceed the longest partial reconfiguration period of one PLD. As was mentioned in Section 7.6, the period of partial reconfiguration could be much shorter than the period of complete PLD reconfiguration and can be in the range of microseconds. Thus, this scheme could be an effective solution for multi-PLD RCS architectures when the shortest mode-switching time is a requirement.

The decentralized (distributed) partial reconfiguration scheme can be implemented in the same way as for the complete multiboot reconfiguration presented in Figure 8.12. In this case, all local controller-loaders must perform both: complete initial configuration of the respective PLD and partial reconfiguration procedures for mode switching. However, in contrast with the multiboot scheme discussed in Section 8.11, those controller-loaders can be allocated on-chip in the form of internal controller-loader (ICL). This approach was discussed in Section 7.7. The configuration memory modules dedicated for PLDs can be considered as configuration cache modules. The configuration cache also could be internal (L1 deployed in internal block-RAM modules) or external (L2 deployed in dual-ported SRAM or DDR-SDRAM), as depicted in Figure 7.10.

The scheme utilizing the external and internal cache modules with internal configuration control is presented in Figure 8.15. This scheme of runtime partial reconfiguration looks similar to the multiboot self-reconfiguration scheme (shown in Figure 8.11) but does not require specific *warm* reconfiguration circuits built in to the PLD microarchitecture. Instead, the initial configuration is needed for all PLDs deployed at the system level of the FCR. During this initial configuration, all necessary runtime reconfiguration circuits should be loaded to each of the PLDs as well as internal cache

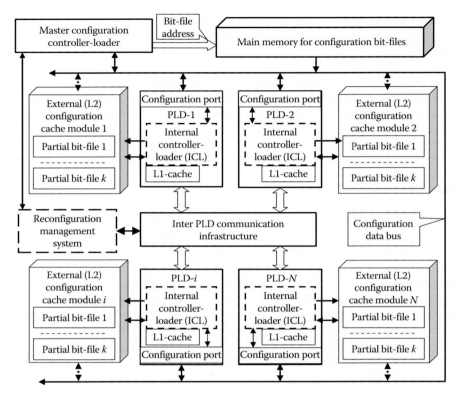

FIGURE 8.15
Distributed runtime partial reconfiguration scheme based on configuration cache.

modules (L1) and external cache memory drivers and ports. The initial configuration process (start-up configuration) can be performed using one of the
earlier schemes presented in Sections from 8.3 through 8.7. As an example,
the parallel passive configuration scheme can be used for the start-up configuration. This scheme was discussed in Section 8.6 and depicted in Figure
8.7. The initial configuration bit-file in this case should be loaded to PLDs
one after another from the main memory containing all configuration files:
(1) initial for complete PLD configuration and (2) all partial bit-files required
for each mode. The initial bit-files must contain configuration data for all
circuits associated with the self-reconfiguration control (e.g., ICL), internal cache organization, and external cache interfacing/driver components.
Further runtime reconfiguration of each PLD is delegated to the individual
internal configuration controllers (ICLs) that can communicate with the RMS
via the inter-PLD communication infrastructure, as shown in Figure 8.15.
Alternatively, communication between ICLs and RMS can be organized
directly by individual PLD-to-RMS channels.

The aforementioned scheme shows that each external cache module contains a certain set of partial configuration bit-files. These files can be loaded to each cache module by the master controller-loader according to the potentially possible mode of operation of the associated PLD and then could be requested by PLD immediately when the bit-file is needed. Since the information of the upcoming mode can be distributed to all PLDs in the system, each of PLDs will address its own bit-file associated with this mode. Obviously, the self-reconfiguration processes will run simultaneously for all PLDs involved in the mode switching process. Significant reduction in reconfiguration/mode-switching time is made possible by this mechanism. However, practical implementation of this scheme requires dual-ported SRAM devices for the external cache modules. In cases where DDR-SDRAM devices are a preferable solution, the partial bit-file loading and updating process could be organized over the ICL receiving the bit-file data via common configuration data bus under control of the master controller-loader.

There can be a number of variations of the aforementioned scheme but all schemes exploiting dedicated external/internal cache modules and partial self-reconfiguration under the control of the internal controller-loader may provide the fastest switching from one mode of operation to another.

This scheme can be considered as one of the most efficient for multitask and multimodal stream processing applications requiring a rapid response to workload or environmental variations.

8.15 Summary

The chapter focused on two major aspects of configuration and reconfiguration of RCS resources at a system level:

1. Organization of initial configuration of PLDs at start-up time
2. Runtime reconfiguration of RCS resources in multi-PLD FCR

For both previously mentioned aspects, classification of methods and configuration schemes was given. It is difficult or even impossible to observe all possible configuration and reconfiguration schemes and their possible combinations. Therefore, the most popular and practically useful schemes and methods were presented. The benefits and drawbacks of each type of configuration scheme were discussed, as well as the recommended architectural solutions.

It was shown how the response time and mode-switching periods can depend on the type of initial configuration and runtime reconfiguration schemes. Also, aspects associated with reliability versus complexity of the configuration schemes were discussed in detail. Therefore, one can see how efficiency and survivability of the entire RCS significantly depend on the proper selection of the configuration and runtime reconfiguration schemes for a particular RCS architecture and its application specifics.

Exercises and Problems

Exercises

1. Is the initial configuration for all PLDs necessary at the start-up time of the RCS?
2. If configuration of all PLDs in the system is done only once at start-up time, can this system be considered an RCS?
3. List all possible methods for the initial configuration of multiple PLDs in RCS. What method is (a) the most economical but slowest and (b) the fastest but most expensive?
4. What is the difference between master (active) mode and passive mode of the daisy-chain configuration method for multiple PLDs?
5. Why does the serial ganged configuration scheme provide a shorter start-up time in comparison to the serial daisy-chain scheme?
6. If the system consists of six identical PLDs and each PLD should have the same configuration as others, what acceleration rate can the ganged configuration scheme provide compared to the serial daisy-chain scheme?
7. Which configuration method can be used for multi-PLD RCS for start-up configuration in cases where (a) the daisy-chain serial configuration method is too slow for start-up requirements and (b) each PLD has its original configuration bit-file?
8. What is a more economical solution from the hardware point of view: (a) parallel daisy-chain configuration scheme or (b) passive parallel configuration scheme?
9. When can minimization of start-up configuration time for multi-PLD RCS be reached using parallel multibus configuration scheme for the smallest PLDs in the family? What benefits and drawbacks are associated with this solution?
10. Why is it necessary to have preconfigured PLDs (with zero configuration start-up time) in the RCS architecture?
11. If there is no preconfigured PLD in RCS, recommend the solution for the start-up configuration of multi-PLD RCS.
12. Why is the runtime PLD reconfiguration needed for the RCS?
13. What are two main classes of runtime partial reconfiguration mechanisms at all levels of the FCR in the RCS?

14. When is the multiboot runtime reconfiguration needed?
15. Is it possible to implement the multiboot runtime self-reconfiguration using any type of PLDs?
16. What kind of hardware support should be included in PLD to perform multiboot runtime self-reconfiguration?
17. When is the reconfiguration management system needed in the RCS architecture?
18. What is the difference between multiboot runtime self-reconfiguration scheme and multiboot runtime reconfiguration scheme with distributed external control? When should the last scheme be implemented?
19. What are pros and cons of a multiboot runtime reconfiguration scheme with centralized external control in comparison with multiboot runtime reconfiguration scheme with distributed external control?
20. In which case should the multiboot runtime reconfiguration be replaced by on-chip partial reconfiguration mechanisms?
21. What is the main requirement for the hardware support of the on-chip runtime reconfiguration scheme?
22. List the benefits and drawbacks for the partial runtime reconfiguration scheme with centralized external control?
23. Describe the cases where use of distributed control in partial runtime reconfiguration scheme is more effective solution than centralized control utilized in the same partial runtime reconfiguration scheme?
24. Why may the distributed runtime partial reconfiguration scheme require on-chip (level 1) and external (level 2) caches for configuration bit-files targeted to PRRs in their respective PLDs?

Problem

1. The given application requires for implementation the following resources: (a) 46,420 LUT-based logic slices; (b) 128 DSP slices, and (c) embedded SRAM with total volume equal to 8MB. The implementation of this application can be done in one of two variants: (i) The single (largest in family) Xilinx Virtex-5 XC5VLX330 FPGA device which can provide: 51840 LUT-slices, 192 DSP slices and 10,368 KB in embedded SRAM blocks or (ii) Ten (smallest in family) Xilinx Virtex-5 XC5VLX30 FPGA devices each of which provide: 4800 LUT-slices, 32 DSP slices 1152 KB SRAM blocks.

    Determine the best variant of implementation that provides the shortest start-up time using the select MAP-32 configuration port working at 66 MHz configuration bus frequency if the configuration bit-file volume for (a) XC5VLX330 is equal to 79,704,832 bits and for (b) XC5VLX30 is equal to 8,374,016 bits.

    Calculate the shortest start-up configuration time for this architectural solution.

References

1. Xilinx Inc., *7 Series FPGAs Configuration*, User guide, UG470 (v 1.10), June 24, 2015, Chapter 9, pp. 153–156 and Chapter 10, pp. 163–176.
2. Altera Corp., *Stratix V Device Handbook, Volume 1: Device Interfaces and Integration*, 2015, Chapter 8, pp. 17–32.
3. Xilinx Inc., *7 Series FPGAs Configuration*, User guide, UG470 (v 1.10), June 24, 2015, Chapter 2, pp. 40–50, Chapter 9, pp. 157–162.
4. Altera Corp., *Stratix V Device Handbook, Volume 1: Device Interfaces and Integration*, 2015, Chapter 8, pp. 4–16.
5. Xilinx Inc., XC2C256 CoolRunner-II CPLD, Data sheet, DS094 (v 3.2), March 8, 2007.
6. Xilinx Inc., Spartan-3AN FPGA Family Data Sheet, DS557, June 12, 2014.
7. Lattice Semiconductor, MachX03 Family Data Sheet, DS1047, March 2015.
8. Xilinx Inc., *Spartan-6 FPGA Configuration*, User guide, UG380 (v 1.0), June 24, 2009, Chapter 7, pp. 115–121.
9. Xilinx Inc., *Virtex-6 FPGA Configuration*, User guide, UG360 (v 3.8), August 28, 2014, Chapter 8, pp. 147–148.

9

Virtualization of the Architectural Components of a System-on-Chip

9.1 Introduction

The general description of the data-computation process and its correspondence to the computing architecture was given in Chapter 1, where a task was defined as an *information object* consisting of two major parts: *task algorithm and data structure*. As per Section 2.4, the *mode of operation* was defined as a combination of (1) one possible variant of a task algorithm, (2) a variant of data structure, and (3) an associated set of constraints (e.g., timing, available resources, power consumption limit). Therefore, the task can be considered as a set of functions, each of which can be initiated in different moments of time according to the initiation conditions. These functions associated with respective data structures can be considered as segments of the task to be *processed without interruption*. The execution of each segment of the task in real-time systems may have certain requirements, including data-processing rate, response time, available resources, and/or power. In turn, each task segment being initiated in different modes may have different implementations in data-processing circuits. In case of relatively low-performance constraints, it may be executed on the instruction-based sequential processor(s) (e.g., arithmetic logic unit [ALU] based). In cases when the high data rate is a must, the dedicated hardware accelerators should be used for this segment implementation. Certainly, it would be difficult to keep all the aforementioned architectural components in real physical form. However, it may be possible to keep them in virtual form as program subroutines for ALU-based implementations or as configuration data-files for implementation as dedicated hardware accelerators.

The aspects associated with task segmentation and creation of virtual components (VCs) accommodating functional segments of the task will be considered in detail in this chapter.

9.2 General Organization of the Task Execution Process

Let us consider the task T as the set of functions $\{F\}$ and associated data structures $\{DS\}$, assuming that each function of the set F_i, $i = 1, 2,\ldots,m$ is directly associated with the respective data structure DS_i and to be initiated by predetermined conditions C_{init_i}.

In addition to the aforementioned, each function may have a specification for performance characteristics such as data-file processing time (e.g., packet or video-frame processing time), in/out data rate, and power consumption limit. These performance requirements can be considered as constraints. For example, in the case of video or image processing, a video data from the entire frame or row of pixels must be executed within the time according to the given video standard (e.g., XGA or 1080 p 60 fps). The same is true for network datagram processing or DSP. On the other hand, there are data and control dependencies between the functions F_i and F_j, $i, j = 1, 2,\ldots,m$, of the task. These dependencies can be represented by a flow chart or any other convenient way.

An example of the task representation in a form of a flow chart is shown in Figure 9.1. In this figure, each function F_i is associated with the data structure DS_i: $F_i \rightarrow DS_i$, where $i = 1, 2,\ldots,m$. The data structure can be (1) digits stored in registers and (2) arrays of data stored in memory module or buffered data-files, direct data-streams received from sensors, etc.

Initial routine F_1 performs all start-up operations and checks the current workload/environment conditions. Then, the mode of operation is selected.

Initially, one of the functions (F_2 or F_3) can be initiated. However, the mode of operation depends on the current set of constraints associated with the selected functionality as well as the type and status of the data source. For example, if F_2 is a video CoDec (e.g., MPEG type), the mode of operation depends on MPEG-2 or MPEG-4 or other particular types of video-compression algorithm. At the same time, the data structure provided by the video sensor is the video data-stream in one of the possible standards (e.g., 720 p × 30 fps, 1080 p × 60 fps). Each of the combinations may require a different way for the accommodation of the same functionalities and interface/data-buffering specifications. Certainly, MPEG-2 CoDec working with 720 p × 30 fps video standard will have a different hardware implementation in comparison with MPEG-4 CoDec working with 1080 p × 60 fps video-stream standard.

The selected functionality depends on the actual consequence of possible functions according to algorithm and data dependencies. The earlier example shows that if F_2 is selected ($C_{init\_2 = true}$), then the F_4 or F_m could be initiated but not F_3 or F_i. Similar to this, if $C_{init\_3 = true}$, F_3 should be initiated (e.g., JPEG CoDec), and therefore, F_i or F_m could be initiated consequently in the near future. According to the earlier conditions, the following general conclusions can be made:

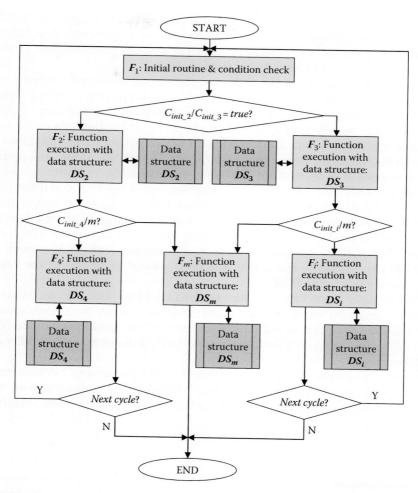

FIGURE 9.1
Example of the task algorithm reflecting data sources associated with functions.

1. If the task algorithm contains alternative functions, not all functions of a task are initiated at a time. Therefore, there are *active and passive functions* at any time of task execution.

2. If the task algorithm contains control and data dependencies between functions, it is possible to predict which passive functions could be *initiated soon*.

3. Each function is associated with own known-ahead data structure. Also, each pair *function and data structure* can be associated with its respective set of required performance constraints.

Hence, it is possible to see that, in general case, only active functions should be in execution stage when passive functions can stay in memory. On the other hand, allocation of passive functions in memory could be done according to their dependencies to the active functions.

This understanding of the task execution process caused memory virtualization in conventional instruction-based computing systems, where an *instruction cache* contains blocks of instructions of active functions or functions to be referenced soon. The same is true for the data cache that contains blocks of data associated with active functions or functions to be referenced in the near future. Obviously, the nature of the task (application) does not depend on the type of processing platform where it should be executed. Therefore, the organization of the task execution process in RCS should reflect the aforementioned specifics of a task nature. In other words, the efficiency of resource utilization could be significantly increased if the FCR accommodates only the active functions. The functions that may be referenced soon in accordance with their control or data dependencies to the active functions should be stored in the cache deployed close to the FCR.

The same is correct for data-files or streams. Only data sources (sensors, buffers, memory modules, etc.) associated with active functions should be activated. The rest of the data sources may stay in sleeping mode. The data-files also could be allocated close to the FCR in order of probability for referencing.

Hence, if not the entire task is loaded to the processing circuits (such as in ASIC or ASIC-like "monolithic" FPGA-based systems), then the task segmentation becomes an important aspect and requires detailed consideration.

9.3 Segmentation of a Task and Concept of Functional Segment

As of the general representation of a task algorithm, an example of which is shown in Figure 9.1, the algorithm consists of a *determined set of functions*, each of which has control and data dependencies with other functions. According to the mathematical definition, *a function* is as a relation between a set of inputs and a set of permissible outputs with the property that each input is related to exactly one output [1]. Each function includes a determined set of *elementary operations*. Elementary operation can be considered as calculation of *one output (result) out of two inputs (operands)*. Arithmetic and logic operations are the most common elementary operations composing a function.

In this consideration, each function in task algorithm can be associated with

1. Set of elementary operations composing the algorithm of the function
2. Determined data structures (e.g., arrays, lists, trees)
3. Set of results generated as the result of function execution during certain period of time
4. Determined set of initiation and termination conditions

The initiation and termination conditions can be (1) conditions satisfied by the results of the previous functions, (2) external (to the task) commands (e.g., from the operating system or sensors), (3) process synchronization signals and flags, etc.

For example, signals like "Enable" or "Reset" can be considered as initiation and termination conditions, respectively, for any function of a task. Examples of process synchronization external signals could be "Video-frame valid" and/or "Row-valid." These signals may also be considered as initiation and termination conditions for video-processing functions.

The set of possible data sources can include, but not limited to, sensors with their interfaces, buffers, and data memory modules. It is assumed that all temporary results of the function execution may be stored in these buffers and memory modules too. Also, the results of the function execution could be single digit, data-file, certain flags or bits in status, or control registers, etc.

Therefore, the smallest segment of a task S_i can be determined as a segment performing a given function: *functional segment*.

Definition 9.1 *Functional segment* FS_i is the part of a task T that consists of (a) the function F_i composed by elementary operations of the functional algorithm, (b) the associated data structure DS_i, and (c) initiation and termination conditions C_{init_i}/C_{term_i}.

The graphical representation of the functional segment FS_i according to the function F_i in Figure 9.1 is depicted in Figure 9.2.

The functional segment can be considered as the *base task segment*. In other words, the execution of a processing cycle of a functional segment cannot be interrupted. As an example, the *elementary functional segment—an instruction* in instruction-based processing systems—can be taken.

Indeed, the ordinary instruction consists of one operation (arithmetic or logic), a data structure consisting of two operands, and a destination of the result. The initiation condition is the instruction address reference and the termination condition is the completion of instruction and program counter update. More complicated functional segments are based on

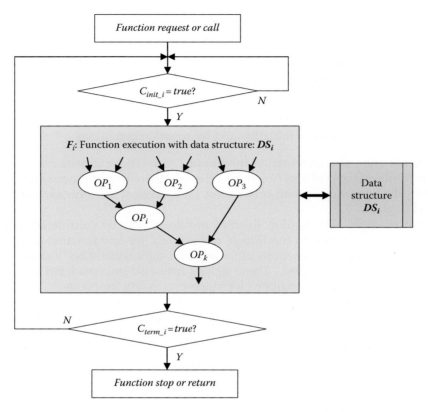

FIGURE 9.2
General structure of the functional segment.

application-specific macrooperations such as DSP functions and video or image processing functions. In all the aforementioned cases, the execution process in the respected functional segment is not interruptible.

Obviously, segmentation of the task can be done in different ways. However, any task segment should consist of one or multiple functional segments: $TS_j = \{FS_i\}$, $i = 1, 2, \ldots, m$; $j = 1, 2, \ldots, n$.

There could be different factors influencing on the selection of task segment granularity:

1. *Specification requirements for a given mode of operation.* For example, application specification requires a noninterruptible performance of the certain mode (e.g., MPEG-2 compression with video-packet transmission over IP). In this case, several functional segments must be grouped in one task segment performing the requested *macrofunction.*

2. *Performance requirements for given mode of operation.* Strict performance constraints may require combining several functional segments in one *macrofunction segment* for further implementation in dedicated macrofunction-specific processing circuit. This aspect will be discussed in detail in Section 9.4.

3. *Mode-switching requirements for a given mode of operation.* In most real-time systems, it is important to provide response time and/or mode-switching time in a specified range. In the case of RCS, this often means available reconfiguration time. Therefore, the "reaction" time of the system to the environmental events or workload changes depends on the volume of configuration bit-file to be loaded to the FCR. In turn, that could limit the granularity of the task segments. This aspect also will be considered in Section 9.4.

4. *Architectural constraints of an existing computing platform.* The architecture organization of the computing platform also can put certain limits on task segmentation. There can be technical limits in memory/cache organization, addressing mechanisms, configuration bus bandwidth (in the RCS), etc.

There could be other reasons that influences the granularity of the task segmentation associated with the organization of computing processes, programming aspects, specifics of compilers or CAD systems, etc. However, in further consideration, only the aforementioned reasons will be discussed.

9.4 Segmentation according to Specification and Performance Requirements

The most common reason for task segmentation according to the mode of operation is just functional specification requirements. In other words, the specification by itself dictates which functions should be combined together to perform the required mode of operation. In the example shown in Figure 9.1, if the mode M_1 requires MPEG-2 video-stream compression and further packet-transmission over the Internet (e.g., IP-TV), the segment of this task may be combined from F_2 (MPEG-2 compression [2]) and F_4 (TCP-IP packets generation [3]). The alternative mode M_2 specification may require MPEG-4 compression for ASI-based transport stream transmission (e.g., MPEG-TS). For this mode, the task segment may include F_3 (MPEG-4 compression [4]) and F_i (ASI-packets generation [5]). An example of such segmentation is shown in Figure 9.3.

It is possible to consider F_2 as the separate segment $TS_1 = \{F_2$ initiated by condition $C_{init\_2}\}$ and store the result of this segment execution in memory.

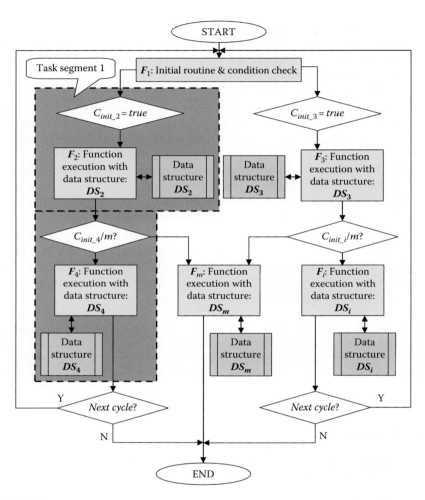

FIGURE 9.3
Segmentation according to the functional specification.

Then, $TS_2 = \{F_4$ initiated by condition $C_{init\_4}\}$ can process the results of the previous segment TS_1 represented in data structure DS_4 in the respective memory unit. However, higher granularity of the task segment allows increasing performance and power efficiency in data-processing. Indeed, more elementary operations are directly communicating to each other and less memory transactions back and forth are needed. This fact allows better utilization of data parallelism and parallelism in algorithm structure. In turn, it enables acceleration of data-execution process as well as minimization of power consumption by processing circuits. This aspect will be detailed in Section 9.8.

The opposite extreme solution is segmentation of the task by atomic functions—elementary arithmetic or logic operations encoded in the form of instructions. In this case, each result must be stored in a memory unit (register or memory cell). Then, one of the next instructions retrieves this result to use it for its elementary operation. Obviously, the amount of operands/results transactions from/to memory unit(s) is proportional to the number of operations and results respective waste of energy and time. To minimize this timing/power overheads, different techniques (such as *forwarding*) have been created and implemented in different RISC architectures [6]. These techniques allow delaying the results of one operation by certain number of clock cycles in the processing circuit to be delivered to the ALU for the next operation avoiding memory transaction. In all cases, longer data-execution pipeline accelerates data-processing rate.

Therefore, *higher granularity of segmentation may result in acceleration of data-execution process* and improve power efficiency. However, the larger task segment will need more FCR resources for its accommodation. In turn, configuration bit-file will be larger accordingly. The larger configuration bit-file will require longer configuration time on the same hardware platform and thus longer mode-switching or reaction time of the system.

Hence, finding the proper balance between the data-execution rate and the mode-switching/system reaction time is one of the main trade-off aspects in the determination of the appropriate segment granularity.

9.5 Segmentation according to Mode-Switching Time and System Constraints

As per Definition 2.12 given in Chapter 2, the mode of operation is one of the possible combinations of the task algorithm, associated data structure, and set of constraints. For example, as presented in Figure 9.1, any change of compression algorithm or video standard (as video data structure) or frame rate (as timing constraints) can be considered as change of the mode of operation. In other words, any change of the workload or environmental change associated with the respective change of system functionality, data structure, or performance specification should be reflected in the mode-switching process. Obviously, this switching process takes certain time that depends on several factors:

1. Architecture organization of the FCR,
2. Volume of configuration bit-file required for the new mode
3. Organization of reconfiguration subsystem and reconfiguration process at all levels of the RCS

It is necessary to mention that in many cases the limit for the mode-switching or system reaction time dictates by the application itself. For example, image processing application working with full HD video standard (1080 p × 60 fps) and requiring mode switching not exceeding one video-frame period (16.66 millisecond) can provide *seamless mode switching*.

Nonetheless, if seamless switching process is not a requirement, application specification may determine the aforementioned mode-switching period. Accordingly, the volume of the configuration bit-file can be defined using Equation 8.1 presented in Section 8.7:

$$T_{conf} = \frac{V_{pld}}{W_{conf\_bus} \times f_{conf}}$$

where V_{pld} is the volume of configuration bit-file when the configuration bus width is W_{conf_bus} and the configuration bus frequency is f_{conf}. These last two parameters are reflecting architecture organization of configuration subsystem of the RCS and reconfiguration process organization.

The reconfiguration time required for switching to the mode $M_i - T_{conf}(M_i)$ can be considered as the mode-switching time $T_{sw}(M_i) = T_{conf}(M_i)$.

Following these arguments, the maximum volume of the associated configuration bit-file to be loaded to the appropriate PLD to switch its functionality to the mode M_i should not exceed

$$V_{pld}(M_i) = T_{conf}(M_i) \times W_{conf\_bus} \times f_{conf} \qquad (9.1)$$

This volume can be taken as the base for the estimation of the possible reconfigurable partition (RP) of the resources available for switching to the required mode of operation. Obviously, different modes may require different mode-switching periods and therefore different limits for RP area.

If the required area exceeds the FCR in one PLD, the parallel reconfiguration of two or more PLDs at the system level of the RCS could be considered as a possible solution.

On the other hand, if the area of the required resources is significantly smaller than the FCR volume in a single PLD, the partial dynamic reconfiguration of this PLD should be taken as an on-chip architectural solution.

For example, considering the aforementioned case of mode switching for video-processing application working with video standard 1080 p 60 fps and requiring mode switching for the next video frame, it is possible to constrain the mode-switching time to 1 second/60 fps = 16.66 millisecond. Selecting the SelectMap-32 as 32-bit configuration port in Xilinx Virtex devices working at configuration bus frequency = 100 MHz, the maximum volume of configuration bit-file is

$$V_{pld}(1080\,\text{p} \times 60\ \text{fps})\,\text{SelectMap-32/100 MHz}$$

$$= 0.01666\,\text{second} \times 32\,\text{bit} \times 100{,}000{,}000\ 1/\text{second} = 53{,}312{,}000\,\text{bits}$$

According to the configuration user guide for Xilinx Virtex-6 FPGA devices [7], this configuration bit-file is enough for complete reconfiguration of Xilinx Virtex-6 FPGA devices up to XC6VLX130T device with total configuration bit-file volume equal to 43,719,776 bits. At the same time, if the largest (in LX series) FPGA is used (XC6VLX760) with total configuration bits equal to 184,823,072 bits, then the RP available for the mode switching will be 53,312, 000/184,823,072 × 100% = 28.8% of the total FCR. Obviously, the organization of configuration scheme and procedure directly influences on the RP area. Taking the aforementioned example and considering the SelectMap-8 working with frequency 50 MHz will reduce the possible configuration bit-file volume to

$$V_{pld}(1080\ \text{p} \times 60\ \text{fps})\ \text{SelectMap-8/50 MHz}$$

$$= 0.01666\ \text{second} \times 8\ \text{bits} \times 50{,}000{,}000\ 1/\text{second} = 6{,}664{,}000\ \text{bits}$$

which is equal to 6,664,000 bits/184,823,072 × 100% = 3.6% of the total FCR available in the Xilinx XC6VLX760 FPGA device.

9.6 Implementation of Functional Segments in the Form of Virtual Components

As mentioned in the previous sections, segmentation of a task may depend on the architecture of the computing platform. In other words, it depends on the variants of physical implementation of the task functions in respective processing circuits. The implementation variant of task functions directly depends on the performance requirements, organization of data (operands) delivery to the processing circuit, and utilization of results (e.g., storage in memory, transfer to the downstream processing circuit(s) or output interface). Since function is considered as the set of elementary operations on data, the respective set of physical operators should be provided in a form of electronic circuits. These electronic circuits could be analog or digital (logic) processing circuits composing analog or digital data processors. In other words, each task function must be "covered" (ensured) by an associated processing circuit. This circuit should include (1) input/output interface to receive operands and transmit results, (2) data-processing unit optimized on given function, and (3) associated memory elements (e.g., registers, buffers, data cache).

Definition 9.2 *Function processing unit* (*FPU*) is the function-specific processing circuit performing (1) data acquisition, (2) data-execution according to the algorithm of a given function, and (3) storage or transmission of the result(s) ensuring required performance characteristics.

An example of an arithmetic FPU is floating-point adders or multipliers that often are added as performance accelerators to conventional ALUs. More complex FPUs are DSP blocks oriented on digital signal processing functions such as FFT, IIR, or FIR.

FPUs are often considered as *performance accelerators* because these circuits can be optimized on algorithm specifics and the data structure of function(s) they are processing. Therefore, the performance of these circuits can be much higher in comparison with general-purpose *multioperational* ALUs. Any multioperational computing circuit can perform only one of the many possible elementary operations at a time. Thus, execution of a complex function consisting of many elementary operations will take a relatively long time because operations can be executed on the multioperational ALU sequentially. In contrast to ALU, function-specific FPU can exploit natural parallelism of the algorithm and parallelism in data structure.

As mentioned in Section 1.1, the data-computing process consists of two flows: (1) flow of data to be processed and (2) flow of control information. Therefore, for any function, it is possible to consider data-level parallelism (DLP) and control-level parallelism. In cases when control information is presented in a form of operational code (op-code) in respective instructions, the instruction-level parallelism can be considered.

1. *Data parallelism* assumes that multiple data-elements can be executed simultaneously. In turn, it means that there are certain sets of independent data-elements (e.g., bits, data-words, vectors of data) that can be processed in parallel.
2. *Control parallelism* assumes that the algorithm consists of sets of elementary operations (algorithm branches) that can be processed in parallel.

Both classes of parallelism are natural (embedded in a function). For example, pixels in a video frame can be processed independently because they are considered as independent (from each other) elements of the picture. Obviously, a real picture consists of many objects, each of which is represented by a set of associated picture elements (pixels). Those picture elements may not be independent in real-life scene. However, in respected video frames, these picture elements *are considered as independent*. The same is true for network packet processing, where data-elements (data-words) are considered as independent. In reality, these data-elements could be elements of a message depending on each other by the sense of this message. Nonetheless, these data-elements on the level of data transmission are considered as independent providing data parallelism in data structure.

When looking at most functional algorithms, it is possible to see that there are branches of operations that could be processed in parallel. For example,

$$Y = \sum_{i=1}^{n} (k \times A_i + l \times B_i) \times C_i, \quad \text{where } i = 1, 2, \ldots, 1024$$

Multiplication operations, that is, $k \times A_i$ and $l \times B_i$, can be processed in parallel since they are independent branches in the function's algorithm. However, other operations like the addition of the aforementioned multiplication results—$k \times A_i + l \times B_i$ or multiplication—$\times C_i$ are sequential in this algorithm due to the order of operations in this function.

At the same time, DLP in data structure of this function allows execution of any set of data: A_i; B_i; C_i for each $i = 1, \ldots, 1024$ in parallel.

It means that function processing time can be significantly accelerated if 1024 processing circuits could execute 1024 data-elements simultaneously. Here, each data-element is a vector:

$$\{A_i, B_i, C_i\} \quad \text{where } i = 1, 2, \ldots, 1024$$

In other words, acceleration of function processing may depend on the number of independent data vectors and the amount of resources allocated for the function execution.

The longest processing period of this function will be if only one general-purpose ALU is used for the function execution. However, the highest performance can be reached when the array of FPUs optimized for function's algorithm are allocated for function processing. Obviously, the hardware cost and the power consumption are in trade-off to performance acceleration.

In other words, there is a *large variety of FPU implementations* providing different performance characteristics vs. cost, area, and power consumption associated with each particular variant of the FPU implementation.

From the opposite point of view, it is possible to say that for each set of performance requirements and system constraints, the appropriate variant of **FPU_j** $j = 1, 2, \ldots, k$ can be found:

$$\{\text{Performance constraints \#1}\} \rightarrow \boldsymbol{FPU_1}$$

$$\cdots\cdots\cdots\cdots\cdots\cdots\cdots\cdots\cdots\cdots\cdots\cdots$$

$$\{\text{Performance constraints \# } j\} \rightarrow \boldsymbol{FPU_j}$$

$$\cdots\cdots\cdots\cdots\cdots\cdots\cdots\cdots\cdots\cdots\cdots\cdots$$

$$\{\text{Performance constraints \# } k\} \rightarrow \boldsymbol{FPU_k}$$

Hence, if the performance requirements of system constraints are changing in time, the appropriate change in variant of FPU implementation allows adaptation of system to the aforementioned variations of performance, workload, or system parameters. Obviously, if all possible FPU variants are implemented in a form of electronic (e.g., logic) circuits, the system may become very expensive and nonefficient from a resource utilization point of view. Accordingly, such solution will significantly increase power consumption. However, if FPU is encoded in configuration bit-file to be programmed in the FCR when needed, the cost-efficiency as well as power efficiency can be kept in a relatively high level. This will enable the RCS to be dynamically adaptive to the environmental or workload variations. Hence, the effectiveness of the RCS and its adaptivity can be significantly increased if all FPUs are represented in the form of a code to be loaded to the FCR. This method will allow configuring the requested FPU when needed and only for the period of respective function processing. Then, this area in the FCR can be used for other FPUs requested by a task and according to the task algorithm. From a hardware point of view, instead of physical components (actual FPU circuits), *VCs* can be configured on-chip. These VCs can appear and disappear in the FCR according to the task requirements, environmental conditions, or changes in system constraints.

Definition 9.3 Virtual component (VC) is the information object (code) which represents procedural and/or structural organization of the FPU architecture optimized for processing a given function on FCR ensuring required performance and system constraints.

There are three possible ways to represent VCs in the system:

1. Virtual software component (VSC)
2. Virtual hardware component (VHC)
3. Virtual hybrid component (VHbC)

Definition 9.4 *Virtual software component* (*VSC*) is a VC represented by an instruction-based code to be executed on multioperational ALU-based computing circuit.

The software component represents only the procedural part of the FPU architecture encoded in a form of a sequential program to be executed on the ALU-based processing core. An example of the VSC can be any program module (e.g., subroutine) that can be called for execution on the processor with a determined set of elementary operations implemented in its instruction set.

The processing core can be implemented in one of two possible forms:

1. *Hard-core processor*—implemented in actual hardware (logic) circuits

2. *Soft-core processor*—configured in the FCR on the base of CLBs and RAM blocks (e.g., MicroBlaze in Xilinx FPGAs [8] or NIOS in Altera FPGAs [9])

Definition 9.5 *Virtual hardware component* (*VHC*) is a VC represented by a configuration bit-file to be loaded to the FCR and configure the function-specific logic circuit.

VHC represents structural organization of the FPU architecture (hardware components and links) and is encoded in a form of configuration bit-file to configure the FPU in the FCR when loaded.

Definition 9.6 *Virtual hybrid component* (*VHbC*) is a VC represented by a combination of procedural and structural parts of the FPU and encoded in a form of (a) a sequential program to run on soft-core processor and (b) a configuration bit-file of associated hardware accelerating circuits.

The determination of the form of FPU implementation depends on performance requirements and ability to use data or control parallelism in a given function and its data structure.

In cases when the function is algorithmic intensive and/or there are no strict performance requirements, the FPU can be implemented as VSC.

In cases when function is computation intensive and there are strict constraints for performance, the implementation of FPU may have no other choice than implementation in the form of VHC.

The virtual hybrid component may be a solution in cases when some computation-intensive parts are combined with algorithmic-intensive parts of the function.

This decision process is similar to the software/hardware codesign process and will not be elaborated here in details. Nonetheless, possible ways for computation acceleration in the FPU will be discussed in the next sections due to the importance of this aspect for FPU implementation in cases when the software implementation does not satisfy the performance constraints.

9.7 Computation Acceleration Exploiting Different Sources of Parallelism

The major source for the acceleration of data-execution process for the past few decades was the increase of clock frequency from units of megahertz to units of gigahertz. This source was based on respective improvements in CMOS technologies when reduced dimensions of MOS transistors allowed a significant reduction in gate capacitance and associated reduction in transistors switching time. However, when the aforementioned switching time reached a subnanosecond range, other factors such as reduced signal propagation distances on-chip, increased leakage current levels, and significant power dissipation in smaller areas have limited the effectiveness of this factor for further increase of system performance. In other words, many factors start to bring their importance for further acceleration of sequentially organized data-computing process. Therefore, one of the main sources for computation acceleration became data and control parallelisms in a given functional segment of a task. Thus, this aspect should be considered in light of the development of VCs and their ability for utilization of both aforementioned types of parallelism.

Exploiting *the data parallelism* assumes that the data structure of the functional segment consists of *independent data-elements*, each of which can be processed simultaneously with any other data-element. There are two ways of processing independent data-elements: (1) in temporal domain and (2) in spatial domain.

Execution of independent data-elements in temporal domain assumes utilization of pipelined circuits. Alternative solution, that is, execution of data-elements in spatial domain, requires execution of multiple data-elements in parallel on the array of processing units. First, the pipelined processing circuits will be considered as one of the most effective way for exploiting the data parallelism.

9.8 Computation Acceleration Using Pipelined Processing Circuits

The pipeline organization of a data processor assumes that operational units (operators) OU_i associated with respected elementary operations O_j, $j = 1, 2, \ldots, n$ are connected to each other in a form of chain, when result of the previous operator can be used as operand for the next operator(s). The general organization of the pipelined processing circuit is depicted in Figure 9.4.

Let us consider the data-execution process on the pipelined processing circuit for the case when the functional segment *FS* consists of a set of

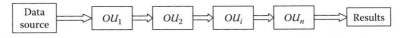

FIGURE 9.4
General organization of the pipelined processing circuits.

elementary operations $\{O_i\}$, $i=1, 2,\ldots,n$ and associated data structure and **DS** consists of the set of independent data-elements $\{D_j\}$, $j=1, 2,\ldots,m$.

If the process of data-execution of the aforementioned functional segment is organized in a form of pipeline, each data-element (e.g., vector of data-words) should path all operational units from the data source to the end of the pipelined data-path. Thus, after getting the first result, when the last (in pipeline) operational unit (OU_n) is processing the data-element D_j, the previous operational unit (OU_{n-1}) will perform its operation on the data-element D_{j+1} and OU_{n-2} will process D_{j+2}.

An example of data-execution process on the simple pipeline is shown in Figure 9.5.

In this example, all operations require the same period of processing time T_{cycle}. In this example, the data structure consists of the array of five independent data-elements, each of which should path four elementary operations (e.g., arithmetic, logic, or storage operations). The data-element could be a data word or set of words (vector). For example, a pixel in a color video frame can be considered as a set of three words associated with red, green, and blue colors.

Figure 9.5 shows that the operational unit OU_1 starts processing the next data-element as soon as it completes its operation on the previous data-element, the result of which is transferred to the next operational unit. Thus, after four time cycles (operational cycle periods), the first result is ready for output. Then, each next result is generated at the end of every cycle time.

The period of time from the initiation of data-execution period until the generation of the first result is called "latency period." Further periods of results generation are called "cycle times."

Figure 9.5 shows that starting from the last operation O_4 on the first data-element D_1, all operational units are working in parallel, executing, however, different data-elements within the same cycle time. In turn, this allows acceleration of data-processing up to four times, which is the number of operational units working in parallel.

Data element	Time Cycles								
	# 1	# 2	# 3	# 4	# 5	# 6	# 7	# 8	# 9
D_1	O_1	O_2	O_3	O_4					
D_2		O_1	O_2	O_3	O_4				
D_3			O_1	O_2	O_3	O_4			
D_4				O_1	O_2	O_3	O_4		
D_5					O_1	O_2	O_3	O_4	
Results					Result 1	Result 2	Result 3	Result 4	Result 5
Timing elements	Latency period				Cycle time 1	Cycle time 2	Cycle time 3	Cycle time 4	Cycle time 5

FIGURE 9.5
Example of a pipelined data-execution process.

Let us analyze the speedup of the pipelined processing circuits in general.

If the number of operations O_i is n and the number of independent data-elements to be executed D_j is m, then the time for execution of all data-elements on the pipelined processing circuit $T_{exe\_pipeline}$ will be equal to

$$T_{exe\_pipeline} = Latency\ period + (m-1) \times Cycle\ time \qquad (9.2)$$

in cases when the period of any elementary operation is adjusted to *Cycle time.*

The latency period is equal to the sum of all operation times included in the pipeline in sequence.

Thus, latency period for this "ideal" pipeline is equal to *Latency period* = $n \times Cycle\ time.$

Therefore, the $T_{exe\_pipeline} = n \times Cycle\ time + (m-1) \times Cycle\ time = Cycle\ time \times (n+m-1).$

In cases when each data-element is executed on the nonpipelined (sequential) processing circuit (e.g., ALU based), the processing time for each data-element will be equal to the sum of operational times of all operations required by algorithm of the functional segment. Assuming the previous assumption that all operational times are equal to the cycle time, the execution time for all data-elements on the sequential processing circuit is equal to $T_{exe\_sequential} = m \times n \times Cycle\ time$

Thus, the speedup of the pipelined circuit in comparison with the sequential processor will be equal to

$$Speedup\ (pipeline/sequential) = \frac{m \times n \times Cycle\ time}{(n+m-1) \times Cycle\ time} = \frac{m \times n}{n+m-1} \qquad (9.3)$$

This equation shows that speedup of the pipelined circuit does not depend on cycle time and depends only on values of m and n.

Let us consider the case when $m \gg n \gg 1$. In this case, Speedup = $n \times m/m = n$.

In other words, when the number of data-elements is much higher than the number of operations, the speedup of the pipelined processing circuit is equal to the number of operations or number of operational units working in parallel.

On the other hand, when $m = 1$, the speedup = $1 \times n/(n+1-1) = 1$. Thus, there is no speedup of computation process when there is only one *independent* data-element. Between these two extremes, there are increasing values of the speedup from 1 to n.

In Table 9.1, an example of the aforementioned computation accelerations (speedup) is calculated for different numbers of data-elements for the functional algorithm consisting of 10 operations.

According to this table, the speedup value increases in respect to the number of data-elements to be processed but never exceeds the number of operations.

TABLE 9.1

Speedup Provided by Pipelined Processor in Comparison with Sequential
Processor

Number of data-elements	1	2	4	8	16	32	64	128	256
Speedup for the pipeline with 10 stages	1	1.818	3.077	4.705	6.4	7.805	8.767	9.343	9.66

Hence, the pipelining circuit is effective in terms of computation acceleration when the number of independent data-elements in the data structure of
the functional segment is much higher than number of operations included
in this functional segment. On the other hand, the highest acceleration never
exceeds the number of operations in this segment.

In sum, more operations being pipelined allow higher acceleration of data-
execution in the case of large massive independent data-elements and cause
longer latency period. Alternatively, when the number of independent data-
elements is relatively small, the pipelined organization of data-execution
process may not be an effective solution. Thus, the data structure and execution time limit are defining the computation-intensive functional segment
from other segments of the task.

Let us consider now the case when operation period is not the same for different operations included in algorithm of the functional segment. For example, the data-execution period for elementary arithmetic and logic operations
(e.g., ADD, NAND, XOR) may be shorter than the execution period required
for integer multiplication or division operations. In this case,

$$Latency\ period = \sum_{i=1}^{n} T_{Oi} \qquad (9.4)$$

where T_{Oi} is the execution period of operation O_i, $i = 1, 2,\ldots,n$ on one data-
element. The cycle time, however, will be equal to the longest execution
period required for one or more operational units. Since the next stage
(operational unit) in the pipeline cannot start its operation before getting the
result from the previous one, all other operational units must wait for the
completion of longest operation in line. This aspect of the pipeline performance is illustrated on the example in Figure 9.6.

The timing diagram of pipelined operations shown in Figure 9.6 demon-
strates that (1) the latency period is extended due to longer O_3 operation time
and (2) the cycle time became equal to O_3 operation period. Indeed, while
the operation unit OU_3 performs the operation on data-element D_1, it cannot receive the result of O_2 operation on data-element D_2 ready for transferring from the operational unit OU_2. Therefore, OU_2 must wait until the result
is received by the next operational unit O_3. Consequently, O_2 cannot take

Data element	Time Cycles								
	# 1	# 2	# 3	# 4	# 5	# 6	# 7	# 8	# 9
D_1	O_1	O_2	O_3		O_4				
D_2		O_1	O_2	*Wait*	O_3		O_4		
D_3			O_1	*Wait*	O_2	*Wait*	O_3		O_4
D_4				*Wait*	O_1	*Wait*	O_2	*Wait*	O_3
D_5						*Wait*	O_1	*Wait*	O_2
Results							Result 1		Result 2
Timing elements	*Latency period*						*Cycle time 1*		*Cycle time 2*

FIGURE 9.6
Example of a pipelined data-execution process with different operation periods.

result from the previous (in pipeline) unit OU_1. Hence, the operation which requires longest time for execution causes stall of other operations in the pipeline. Thus, the cycle time becomes equal to the aforementioned longest operational period:

$$Cycle\ time = \max\{T_{Oi}\} \tag{9.5}$$

And the time for execution of all data-elements on the pipelined processing circuit will be equal to

$$T_{exe\_pipeline} = \sum_{i=1}^{n} T_{Oi} + (m-1) \times \max\{T_{Oi}\} \tag{9.6}$$

Certainly, the speedup of this pipeline will be less than the ideal pipelined processor due to number of stall (waiting) periods of some operational units during every cycle time. Thus, the main point in designing the pipelined processing circuits is to find a proper set of operational units performing their operations in equal periods. If this goal is reached, the cost-efficiency in resource utilization will be highest possible.

9.9 Computation Acceleration Exploiting Control-Flow Parallelism

As determined in the previous section, the acceleration of computation process utilizing data parallelism in temporal domain (pipelined processing circuits) has a limit equal to the number of operations in the pipeline. Further, acceleration can be reached utilizing control parallelism in algorithm structure or algorithm branch parallelism.

Let us continue the analysis of the example of functional segment discussed in Section 9.6:

$$Y = \sum_{i=1}^{n}(k \times A_i + l \times B_i) \times C_i, \quad \text{where } i = 1, 2, \ldots, 1024$$

The algorithm of this function can be represented in a form of a sequencing graph (SG) according to the methodology of architectural synthesis and optimization of digital circuit [10]. This SG is depicted in Figure 9.7.

The SG model of a computation process is an extended dataflow graph (DFG) since it should represent the flow of data and control sequences for the respective functional algorithm.

SG is a polar and acyclic graph where the first and last vertices are no-operation (NOP) representing initiation and operation control stages of computation process and all other vertices are in one-to-one correspondence with operations in the functional algorithm.

The edges of the SG are representing data dependencies according to the dataflow in this algorithm. The main difference between generic DFG and SG is that SG is acyclic as per definition [10]. This property of the SG allows considering this model as the model of computation in temporal domain.

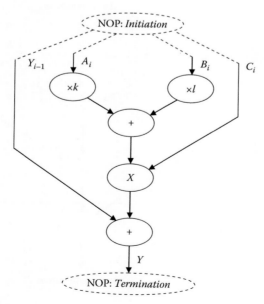

FIGURE 9.7
Sequencing graph of the function $Y = \sum_{i=1}^{n}(k \times A_i + l \times B_i) \times C_i$.

In other words, sequencing of operations could be considered in time, and therefore, SG can be scheduled. In other words, in certain moments of time, the operations associated with data ready for execution can be started. For example, initiation of the functional segment assumes that input data-element consisting of the A_i, B_i, and C_i is available for processing. It is also assumed that initial result of the "previous" operation is $Y_{i-1}=0$. The period of time between T_1 and T_0 is used to latch the values of A_i, B_i, C_i and Y_{i-1} in the internal data-registers of the function processing circuits.

Then, the two operations $(A_i \times k)$ and $(B_i \times l)$ can be initiated at start time T_1 and executed in parallel. However, further operations must wait for the completion of these operations due to data dependencies. When the aforementioned operations are completed, the next operation is performed: ADD the results of $(A_i \times k)$ and $(B_i \times l)$ that can start at T_2. T_3 is a start time of the final multiplication on C_i and the accumulation of the results of each iteration $i=1, 2,...,n$ starts at T_4. Completion of i-iteration is a start time for the termination of the processing cycle. It is necessary to mention that initiation of the next cycle is not represented in the SG because readiness of the next data-element may depend on factors external to the data-execution process represented in this SG.

The scheduled SG representing the aforementioned functional segment is shown in Figure 9.8. The period of time from the T_0 and T_5 represents the latency period from initiation of the execution of data-element to the termination of current iteration. In general case, operation periods may not be equal

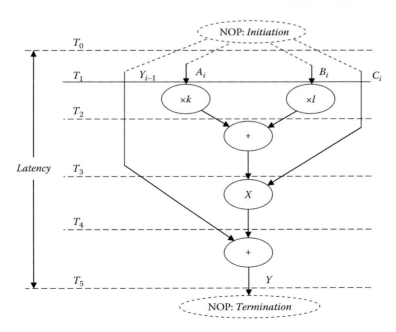

FIGURE 9.8
Scheduled SG of the function $Y = \Sigma_{i=1}^{n}(k \times A_i + l \times B_i) \times C_i$.

to each other. For example, the period of multiplication (e.g., $T_2 - T_1$) most probably is longer than the period of ADD operation (e.g., $T_3 - T_2$). Obviously, the operational periods are depending on the resources used for each type of operation. In other words, there is certain correspondence between coverage of operations by *operational elements*.

Definition 9.7 *Operational element* is a hardware data-processing circuit conforming the determined arithmetic or logic operation.

For example, multiplier is an operational element conforming multiplication operation and adder is an operational element conforming the summation of two digits. Obviously, there could be many possible ways of the implementation of an operational element performing the same type of operations. These different variants of operational element can provide different processing times, power consumption, area of utilized resources, etc. On the other hand, the execution time for one data-element according to the algorithm of the functional segment depends also on the number of operational elements deployed in the dedicated processing circuit designed for given functional segment. In the example discussed earlier, utilization of two multipliers instead of one for two parallel branches of the aforementioned algorithm will reduce the latency and the entire processing time of all data-elements. However, if more than two multipliers are deployed for this functional segment, no further reduction for latency can be gained. The same is true for the adders. Since all ADD operations are in sequence, the latency will not be reduced if more than one adder is allocated for this functional segment. However, the total processing time depends not only from the latency but on cycle time too. And additional operational elements can play significant role in further acceleration of the execution process.

Here, we are coming to the resource binding problem in digital circuits' design and associated correspondence between resource allocation and performance characteristics such as latency and cycle time. This aspect will be considered in details in the next section.

9.10 Computation Acceleration by Proper Resource Binding

The problem of "covering" algorithm operations by their respective operational elements is known as the resource binding problem [11].

Definition 9.8 *Resource binding* is a mapping of each operation represented by the corresponding vertex V_i on the SG to the certain instance of resource type R_i, $i = 1, 2, \ldots, n$.

In other words, the resource binding determines how many instances of each type of the same operational elements should be allocated in the FPU for the functional segment execution.

To illustrate this point, let us consider two extremes of resource binding for the aforementioned scheduled SG: (a) the case of FPU with minimum resources used and (b) the case of FPU with maximum resources allocated for this task segment. If the task segment is represented by the aforementioned function, it requires two types of operation (*add* and *multiply*). Hence, only two types of operational elements are needed. The minimum number of instances of resource is one element per type. In this case, all multiplications must be performed on one multiplier MULT-1 and all summations on one adder ADD-1. Thus, the SG should be scheduled accordingly. This SG is shown in Figure 9.9.

Let us assume that the operation time for the adder is one clock cycle and multiplication takes two clock cycles. In the case of minimum resources (according to the resource binding presented in Figure 9.9), the latency will be equal to eight clock cycles. And the cycle time for the execution of one data-element vector $\{A_i, B_i, C_i\}$ will be equal to seven clock cycles according to the schedule shown in Figure 9.10. Operations in this schedule are numbered as follows:

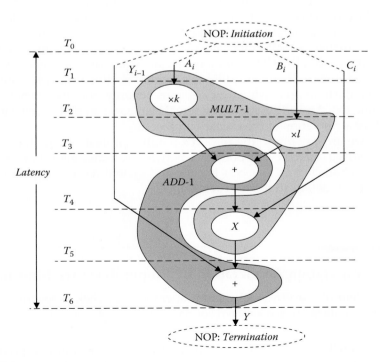

FIGURE 9.9
Scheduled SG with minimum resource binding.

Clock Cycle #	1 2	3 4	5	6 7	8	9	10 11	13	14	15	16	17
MULT-1	[1]*1*	[2]*1*		[4]*1*	[1]2		[2]2		[4]2		[1]3	
ADD-1			[3]*1*		[5]*1*			[3]2			[5]2	
Timing	*Latency* = 8 c.c						*Cycle time* = 7 c.c.					

FIGURE 9.10
Timing schedule for operational elements in case of minimum resources.

$A_i \times k \rightarrow$ [1]*i*; $B_i \times l \rightarrow$ [2]*i*; $(A_i \times k + B_i \times l) \rightarrow$ [3]*i*; $(A_i \times k + B_i \times l) \times Ci \rightarrow$ [4]*i and*
accumulation $(A_i \times k + B_i \times l) \times C_i + Y_{i-1} \rightarrow$ [5]*i*, where $i = 1, 2, \ldots, 1024$

The total execution time of the aforementioned functional segment on the processing circuit containing one adder and one multiplier is equal to

$$T_{exe}(\min) = Latency + (number\ of\ data\text{-}elements - 1) \times Cycle\ time$$

$$= 8\,\text{c.c.} + (1024 - 1) \times 7\ \text{c.c.} = 7169\ \text{c.c.}$$

If the highest performance (minimum processing time) is required, each operation represented by the corresponding vertex V_i on the SG should be one to one *covered* by its individual operational element of respective type. Thus, this circuit will require three multipliers and two adders:

1. MULT-1 for [1]*i* operation, MULT-2 for [2]*i* operation, and MULT-3 for [4]*i* operation
2. ADD-1 for [3]*i* operation and ADD-2 for [5]*i* operation

This allows the creation of fully pipelined processing circuit that provides the shortest latency and highest possible processing data rate. The scheduled SG with maximum resource binding is presented in Figure 9.11.

One of the possible timing schedules for operational elements in the case of maximum resources allocated for this processing circuit is shown in Figure 9.12.

The latency in this case is seven clock cycles and the cycle time is equal to two clock cycles according to Equation 9.5 and the schedule shown in Figure 9.12.

The total execution time of this functional segment on the fully pipelined processing circuit is equal to

$$T_{exe}(\max) = Latency + (number\ of\ data\text{-}elements - 1)$$
$$\times Cycle\ time = 7\ \text{c.c.} + (1024 - 1) \times 2\ \text{c.c.} = 2053\ \text{c.c.}$$

Thus, the speedup provided by processing circuit with maximum operational resources vs. processing circuit with minimum operational resources is equal to

$$\text{Speedup} = T_{exe}(\min)/T_{exe}(\max) = 7169\ \text{c.c.}/2053\ \text{c.c.} \times 100\% = 349.2\%$$

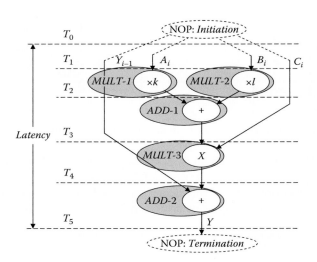

FIGURE 9.11
Scheduled SG with maximum resource binding.

Clock Cycle #	1	2	3	4	5	6	7	8	9	10	11	12
MULT-1	[1]*1*		[1]2		[1]3		[1]4		[1]5		[1]6	
MULT-2	[2]*1*		[2]2		[2]3		[2]4		[2]5		[2]6	
MULT-3					[4] *1*		[4] 2		[4]3		[4]4	
ADD-1			[3]*1*		[3]2		[3]3		[3]4		[3]5	
ADD-2							[5]*1*		[5]2		[5]3	
Results							Y_1		Y_2			Y_3

FIGURE 9.12
Timing schedules for operational elements in case of maximum resources.

It means that additional two multipliers and one additional adder can increase the performance of the FPU based on this dedicated processing circuit almost in 3.5 times. The difference in the performance of FPUs using minimum and maximum hardware resources shows that it is possible to find the configuration of processing circuit with one of the intermediate performances in the range between T_{exe}(min) and T_{exe}(max). Since the algorithm of the functional segment consists of two summation operations and three multiplications, there can be $3 \times 2 = 6$ possible configurations of the number of operational elements in the processing circuit composing respective FPU. Table 9.2 provides all possible configurations of the processing circuit, associated processing times, and speedup in comparison with the processing circuit with minimum operational elements.

Table 9.2 shows that there is no influence of adders to the performance of this FPU circuit. Only the number of multipliers is influencing the speedup

TABLE 9.2

Execution Time according to the Number of Operational Elements

Configuration	1 MULT + 1 ADD	1 MULT + 2 ADDs	2 MULTs + 1 ADD	2 MULTs + 2 ADDs	3 MULTs + 1 ADD	3 MULTs + 2 ADDs
Execution time in clock cycles	7169	7169	4099	4099	2053	2053
Speedup in times		1	1.75	1.75	3.49	3.49

of this FPU. On the other hand, multipliers are more area-consuming and power-consuming operational elements. There is quite proportional dependence between the amount of hardware resources and the FPU performance. However, in some cases, even the highest performance provided by the FPU with fully pipelined processing circuit (data-path) may not satisfy the specification requirements. In this case, parallel processing of segmented data structure can be the only effective solution. This aspect will be considered in the next section.

9.11 Computation Acceleration Using Data Structure Segmentation

In cases when strict constraints for the execution time is a specification requirement, the data structure of the associated functional segment can be segmented for parallel execution on several identical processing circuits working simultaneously. As mentioned in Section 9.8, parallel execution of multiple data-elements on the array of processing units (in spatial domain) can increase the performance of the FPU proportionally to additional hardware resources. This solution can be considered as effective complement to the pipelined data-paths in cases when single data-path cannot satisfy the timing constraints. The main condition for utilization of this solution is the possibility for data structure segmentation on independent segments of data-elements.

If the execution time of a given functional segment on FPU with single pipelined data-path utilizing maximum resources is equal to $T_{exe}(\max)$ and the timing constraint $T_{lim} < T_{exe}(\max)$, the number of parallel data-paths N should be equal or greater than the ratio $T_{exe}(\max)/T_{lim}$:

$$N \geq \frac{T_{exe}(\max)}{T_{lim}} \qquad (9.7)$$

For the example considered in the previous section, the $T_{exe}(\max) = 2053$ c.c. In cases when the specification constraint for the execution the aforementioned functional segment is 2 microsecond and the highest possible clock frequency is 266 MHz, the following estimation of the number of parallel data-paths can be done:

1. The limit in clock cycles is $T_{lim} = 2 \times 10^{-6}$ second $\times 266 \times 10^6$ c.c./ second $= 532$ c.c.
2. The number of parallel data-paths is $N \geq T_{exe}(\max)/T_{lim} = 2053/532 \geq 3.86 = 4$.

In other words, for the aforementioned specification constraints, four identical data-processing circuits working parallel on different segments of the data-file is required. Each of these circuits' data-path consists of three multipliers and one adder (according to Table 9.2) as the most cost-effective solution for the given functional algorithm. One of the possible FPU architecture organizations is presented in Figure 9.13.

In this case, the data structure initially organized as the 1024 data-elements, each of which is a vector of three digits $\{A_i, B_i, C_i\}$ $i = 1, 2, \ldots, 1024$. This is divided into four independent segments, each of which consists of the 256 vectors given earlier. Obviously, it is possible to do if the vectors of data in each substructure are independent from the vectors in any other substructure of data. Thus, if such segmentation of the initial data structure is possible, each of the data segments can be executed separately on the dedicated processing circuit optimized for a given algorithm.

However, at the end of processing all the data-elements, the final result of the function execution needs appropriate integration. In the considered example, the data-integration circuit consists of additional adders accumulating results received from four data-paths. This integration process also requires some additional time (two clock cycles in this case).

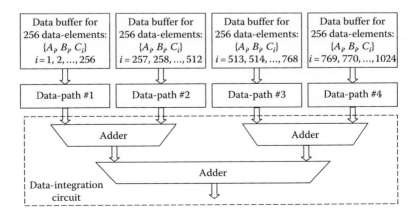

FIGURE 9.13
FPU organization based on four parallel data-paths and data-integration circuit.

Let us estimate the total execution time for the aforementioned FPU. Each data-path consisting of three multipliers and one adder will process 256 data-elements within

$$T_{exe} = Latency + (256-1) \times Cycle\ time = 7\ \text{c.c.} + 255 \times 2\ \text{c.c.} = 517\ \text{c.c.}$$

And integration circuit will add two additional clock cycles for the summation of all four results from each data-path. Therefore, the total execution time will be equal to 519 c.c., which is less than the specification limit equal to 532 c.c.

It is necessary to mention that quite often the exact timing is required. This is the case for video applications, DSP, digital communication, and broadcasting applications and many others. For these cases, the final adjustment can be done by (1) insertion of dummy clock cycles using additional counter or (2) by clock frequency adjustment. In the presented example in the case of dummy cycles, $532-519 = 13$ c.c. must be generated in addition to 519 c.c. spent for data-execution. The 4-bit counter should be added to the FPU control circuit to perform that. Alternatively, the clock generator can be adjusted to generate clock cycles with frequency equal to Fc.c. = 519 c.c./2 microsecond = 259.5 MHz, instead of 266 MHz.

9.12 Organization of the Virtual Hardware Component

In Section 9.6, it was mentioned that a functional segment can be implemented in one of three forms: (1) virtual software component (VSC), (2) virtual hardware component (VHC), and (3) combination of VHC and VSC—virtual hybrid component (VHbC). It was also shown that the form of implementation of the functional segment directly depends on the set of constraints.

The mechanism for the aforementioned implementation is the FPU. The FPU was defined as function-specific processing circuit performing: (1) data receiving and/or buffering, (2) data-execution according to the algorithm of a given function and an associated data structure, and (3) storage/transmission/outputting the result(s).

The FPU must satisfy the performance constraints. In cases when the VSC can provide the required performance, the VSC can be considered as sufficient way for the functional segment implementation. This aspect was discussed in Section 9.6. In cases when VSC cannot satisfy the required performance constraints, VHC or VHbC are possible solutions for FPU implementation. As of Definition 9.1 in Section 9.3, the functional segment FS_i is the part of a task T that consists of (a) function F_i composed by elementary operations of the functional algorithm, (b) associated data structure DS_i, and (c) initiation and termination conditions C_{init_i}/C_{term_i}.

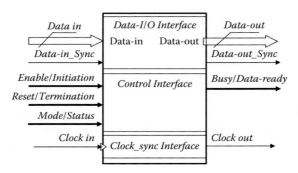

FIGURE 9.14
Symbol of the virtual hardware component.

In cases when functional segment is implemented as VHC, it must include (1) FPU consisting of (a) function-specific data-path, (b) data-structure-specific I/O interface(s), (c) memory elements and (2) control unit performing (a) initiation and termination procedures, (b) synchronization and control signaling for the I/O circuits, and (c) synchronization control signals for data-path to satisfy the required performance, area, and power consumption/dissipation constraints.

The VHC can be represented in a form of component's *symbol*. The general symbol of a component is depicted in Figure 9.14.

The symbol of the VHC must define all the interfaces: (1) data I/O interfaces with associated synchronization/strobe signals; (2) control signaling interface allowing initiation and termination the data-execution process, mode-switching and handshaking signaling, etc; and (3) clock interface for local and global synchronization of the VHC with other components and processes in the system. The symbol also defines the width of data input (data-*in*) and data output (*data-out*) buses. In case of programming of the VHC in hardware description language (HDL) (e.g., VHDL or Verilog), the symbol determines the entity part of the component's hardware description.

The creation of VHC data-path starts from high-level synthesis of digital circuit and optimization similar to any methodology of high-level synthesis used for ASICs (e.g., [11]).

In general terms, this process may be organized as follows:

1. Representation of the data-execution process in the form of SG (SG-VHC)

2. Scheduling and resource binding of the SG to fit the required performance constrains

3. Conversion of the scheduled SG with optimal resource binding to the hardware circuit block diagram

Then, low-level logic design can be performed using an appropriate HDL and associated CAD tools.

Let us briefly consider all the aforementioned stages of VHC data-path synthesis using the example described in Section 9.11. In Table 9.2, two configurations of the data-path provide the highest acceleration rate 3.49 in comparison with the configuration with minimum resources. Both include three multipliers. However, one of the configurations has one adder when the other requires two adders. Since the variant with one adder requires fewer resources, let us consider further this variant. The scheduled SG of this variant is shown in Figure 9.15.

Here, the adder ADD-1 is a shared resource used for two operations allocated in two different time slots. The schedule shown in Figure 9.16 illustrates utilization of ADD-1 for both operations: $(A_i \times k + B_i \times l) \to [3]$ and $[(A_i \times k + B_i \times l) \times C_i + Y_{i-1}] \to [5]$.

This schedule allows utilization of one adder instead of two because the multiplication takes two clock cycles and summation only one. Therefore, the same operational element (adder) can be used twice within multiplication period. However, it requires *additional multiplexors* and *demultiplexor* for this adder because its inputs and outputs should be connected to different inputs and outputs of other devices (multipliers, registers, etc.). The block diagram of this configuration of data-path presented in Figure 9.17

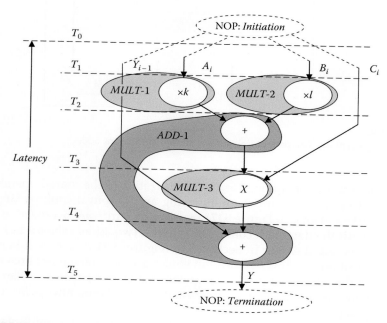

FIGURE 9.15
Scheduled SG with resource binding: three multipliers and one adder.

Clock Cycle #	1	2	3	4	5	6	7	8	9	10	11	12
MULT-1	[1]*1*		[1]2		[1]3		[1]4		[1]5		[1]6	
MULT-2	[2]*1*		[2]2		[2]3		[2]4		[2]5		[2]6	
MULT-3					[4] *1*		[4] *2*		[4]3		[4]4	
ADD-1				[3]*1*	[5]0/[3]2		[5]*1*/[3]3		[5]2/[3]4		[5]3/[3]5	
Results								Y_1		Y_2		Y_3

FIGURE 9.16
Timing schedule for operational elements in case of three multipliers and one adder.

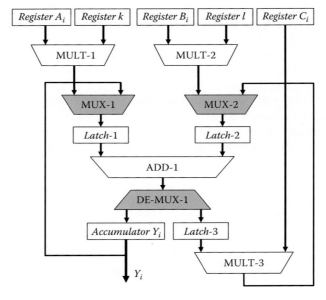

FIGURE 9.17
Block diagram of the data-path utilizing three multipliers and one adder.

illustrates the aforementioned aspect of utilization of the shared operational element. Certainly, the input multiplexing and output demultiplexing circuits (marked in gray in Figure 9.17) are creating hardware overhead in comparison with the data-path configuration with two adders shown in Figure 9.18 and discussed earlier in Section 9.10 (Figures 9.11 and 9.12). It is necessary to mention that this overhead in some cases can take more area in the FCR than the additional dedicated operational element such as adder. This aspect, however, is *hidden in SG representation*, but should always be considered in practice.

Another question arises from timing overhead. The multiplexing/demultiplexing circuits bring additional propagation delays and usually

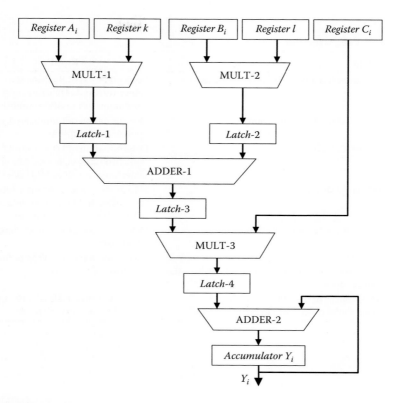

FIGURE 9.18
Block diagram of the data-path utilizing three multipliers and two adders.

require latching the data after transmission over the multiplexor's or demultiplexor's circuits. This may require additional data registers and clock cycle(s) for reliable operation (especially in the FPGA-based SoCs).

Unfortunately, the aforementioned aspects associated with the utilization of time-shared operational elements are not explicitly visible on the scheduled SG with resource binding. However, these aspects always should be kept in mind during the comparative analysis of data-path's configurations.

When the block diagram of the VHC data-path is complete, the low-level logic synthesis of this circuit can be performed together with the associated control unit in a form of finite state machine (FSM). This FSM should provide the schedule determined for the data-path (e.g., shown in Figure 9.16). The "start," "stop," "reset," and mode-switching conditions determined for the VHC are equal to NOP (initiation and termination) vertices on the SG. In other words, these conditions are triggering events for start/stop states of the FSM. This aspect will not be elaborated in this chapter due to well-known methodologies for digital circuit's design (e.g., [12,13]).

TABLE 9.3

Process of VHC Creation

Process Stage	Result
Determination of functional segment specification	Determination of the function algorithm, data structure, and initiation/termination conditions
Determination of the FPU constraints	Determination of timing, area, and power consumption
Creation of VHC symbol	Determination of I/O, control, synchronization, handshaking interfaces, width of I/O buses, etc.
High-level synthesis of the data-path	SG scheduling and resource binding according to FPU constraints and time schedule for resources
Conversion of the scheduled SG to the block diagram of the data-path	Determination of the block diagram data-path
Control unit design according to start/stop/reset and mode-switching conditions and time schedule of the operational elements	Determination of the FSM of the control unit
Low-level synthesis of the FPU circuit and compilation to the configuration bit-file of VHC	Configuration bit-file of VHC to be stored in memory platform

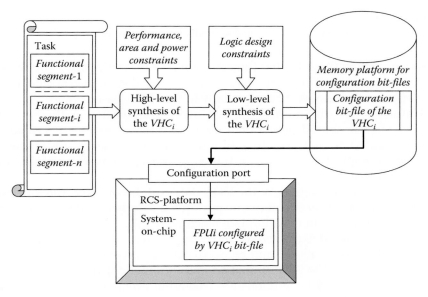

FIGURE 9.19
Process of VHC creation and FPU implementation on the RCS platform.

At the end of the low-level logic synthesis using an HDL and an associated CAD system provided by the vendor of PLD platform (e.g., Xilinx ISE CAD [14] for Xilinx FPGAs or Quartus II CAD for ALTERA FPGAs [15]), the configuration bit-file of the designed VHC is generated. This configuration bit-file being stored in any memory platform (flash card, hard drive or SRAM, etc.) represents the aforementioned hardware component in its virtual form and thus is VHC. This process is illustrated in Table 9.3.

When VHC is loaded to the FCR, it becomes an actual FPU implemented in a form of dedicated hardware circuit and can be incorporated to the system by predetermined procedure associated with FPU interfaces and SoC integration mechanisms. The process from the task segmentation to the incorporation of the FPU to the current SoC architecture is shown in Figure 9.19.

Incorporation of the VHC into the system-on-chip can be static (prior to SoC initiation) or dynamic (in run-time of the SoC). This allows creation of different configurations of data processors and is the subject of the next chapter.

9.13 Summary

Conventional computing systems with fixed architecture allows adaptation of a task to the computing architecture by programming the data-execution procedure (program) and then processing it sequentially (instruction after instruction) on the ALU-based multioperational data-processing circuit. This approach allows keeping different functions and associated data structures in the form of instruction-files and data-files in the instruction and data memories, respectively. Virtualization of memory in modern computing systems enabled significant reduction in computing systems cost and associated cost of data-execution. However, in cases when high-performance, limited power consumption, or limited computing resources are system requirements, the traditional approach based on the concept of computing systems with programmable procedures (CSPP) often cannot be used due to sequential nature of data-execution process on these computers. In such cases, different types of parallelism in data structure or in algorithm must be used. In turn, it dictates the utilization of dedicated hardware accelerators where parallel architecture is optimized for the specifics of the functional algorithm and associated data structure. It is assumed that not all segments of the task require high-performance data-execution but only computation-intensive segments. It is also clear that due to algorithmic and/or data dependencies, not all segments of the task will be active at a time. Instead, only those segments that initiation conditions are satisfied are active at current time. This allows performing task segmentation, considering each *functional segment as a noninterruptible part of the task.* Each functional segment consists of function with associated data structure. Depending on the mode of operation, the functional segment may be required to execute data with given performance requirements. Therefore, several FPUs can be associated with the same functional segment of the task. For example, low-performance implementation may be associated with ALU-based implementation of a given functional segment. However, the high-performance requirements may require function-specific dedicated hardware accelerator. In this case, the software routine running on the instruction-based processing core can be considered as the FPU when in the second case the FPU will be a dedicated hardware circuit configured in the target PLD.

Thus, in this chapter, the concept of virtual processing components was discussed focusing on the organization of the virtual hardware components (VHCs). The concept of VHC assumes that high-performance dedicated data-processing circuit optimized for functional segment and set of performance constraints can be synthesized and then stored in the form of configuration data-file. Then, by the request of the system, this VHC can be loaded to the PLD and become the FPU—physically deployed parallel processing circuit in the system-on-chip for the period of required data-execution. After completion processing of the functional segment, this FPU may be replaced by another FPU. However, the configuration bit-file of this FPU still will be stored in the main configuration memory for further utilization. This approach enables significant reduction in system cost because only those FPUs that are needed for the current mode of operation and current state of a task are implemented as active hardware circuits, the most expensive system resources.

The main problem that can arise in run-time reconfigurable RCS is the incorporation of the aforementioned VHCs into the SoC without violation of the ongoing data-execution process. This problem will be discussed in the next chapter.

Exercises and Problems

Exercises

1. List the reasons that in general case may cause division of task algorithm functions into active and passive, while task execution is performed.
2. What is the role of instruction and data cache in conventional (ALU-based) computers with virtual memory organization?
3. Determine the elementary operation. Which function can be considered as elementary function?

4. List four main elements always associated with the function to be executed.
5. Define the functional segment.
6. What are the components that compose a task segment?
7. Which factors are influencing the task segmentation granularity?
8. What task segment can be considered as macrofunction?
9. Does the number of functions included in the task segment influence (a) the computation acceleration in cases when the task segment is executed on the ALU-based processor? (b) The computation acceleration in cases when the task segment is executed on dedicated processing circuit with pipelined data-path? (c) The mode switching or event reaction time?
10. Which configuration parameters of the PLD(s) should be considered to find the maximum granularity of the partially reconfigurable region (maximum amount of logic resources in PLD) for task segment to satisfy a given mode switching time or event response time of the application?
11. Define the FPU. What are the main components of the FPU?
12. Why is the performance of FPU optimized for the given function much higher than the performance of the conventional multioperation ALU executing the same function?
13. Define the data and control parallelism in algorithm and data structure. How can the data and control parallelism be used for computation acceleration?
14. Why does the way of FPU implementation depend on the performance requirements or constraints?
15. Define the VC. Why does the concept of VC enable significant increase of hardware resource utilization in comparison with the actual physically implemented functional components?
16. What are the possible types of VCs in RCS?
17. What is the difference between virtual software and hardware components?
18. When is the function better implemented in the form of a virtual hardware component and when can the virtual hybrid component become a more effective solution?
19. Why can the pipelined data-path speedup the execution of data structures that consists of independent data-elements?
20. How can the number of operations in pipelined data-path and the number of data-elements in data structure influence speedup?
21. Why does the pipelined data-path cause latency in the generation of the first result and why does the cycle time much shorter than the latency period?
22. What is the difference in cycle time between pipelines with equalized periods of operations included in the data-path and nonequalized pipelines?
23. What represents the vertices and edges on DFG and SG and what is the conceptual difference between SG and DFG?

24. Define the operational element. What is the difference between operational element and functional processing unit?
25. Define "resource binding" process. Why is the reason why proper resource binding on function SG may optimize resource utilization and accelerate function computing process?
26. When is acceleration by data-structure segmentation needed and possible? How does the data-structure segmentation influence the architecture of the respective FPU?
27. What represents the "symbol" of hardware component?
28. List the main parts of the component's interface that should be reflected in the symbol.
29. What are the main stages that require high-level synthesis of the VHC data-path needed before low-level logic circuit design?
30. Why does the utilization of any shared operational element (e.g., adder or multiplier) requires additional hardware overhead as well as timing overhead?
31. Which functions the control unit of VHC should perform (in general case)?
32. List the main stages of VHC creation.

Problems

1. Using the Xilinx Virtex-5 FPGA Product selection Table, perform the following:

    a. Determine the largest FPGA device able to accommodate the FPU to be completely configured in this FPGA within the period of *one video frame*. The FPU processes video-stream in 60 frames per second frame-rate.

    b. Determine the maximum volume of logic resources (number of slices) available for the FPUs allocated in this FPGA device.

    Note: The SelectMAP-32 configuration port can be used when configuration bus frequency is 66 MHz. The complete configuration/reconfiguration time of the target FPGA device *should not exceed one video frame*.

Xilinx Virtex-5 FPGA Product Selection Table

FPGA Device	Number of Slices	Configuration File (Mbit)
XC5VLX30	4,800	8.4
XC5VLX50	7,200	12.6
XC5VLX85	12,960	21.9
XC5VLX110	17,280	29.2
XC5VLX155	24,320	41.1
XC5VLX220	34,560	53.2
XC5VLX330	51,840	79.8

2. Calculate the function processing time on FPU with fully pipelined data-path in cases when the algorithm consists of eight elementary operations: $O_1, \ldots, O_8$.

   The data structure is an array of data-elements that can contain (a) 8 data-elements, (b) 32 data-elements, (c) 256 data-elements, and (d) 1024 data-elements.

   i. Calculate the latency, cycle time, and complete function processing time for each data-structure variant if each elementary operation O_i requires two clock cycles and is performed in order of operation number from $O_1, \ldots, O_8$.

   ii. Estimate the speedup of this FPU in comparison to the sequential multioperational ALU similar to the example presented in Section 9.8.

   iii. If operations O_1, O_2, O_3, and O_4 can be executed simultaneously within the period of two clock cycles, how will the latency and cycle time of the data-execution process change?

   iv. How will the cycle time change in case operation O_5 would require four clock cycles?

3. For the function $Y = \sum_{i=1}^{n} [(A_i + B_i) \times (C_i + D_i)] / 10 \times k$, where $i = 1, 2, \ldots, 2048$:

   a. Create a SG assuming that *NOP* vertex is associated with the initiation and termination of function computation cycle and the rest of vertices are associated with elementary arithmetic operations.

   b. Perform the scheduling of the SG and the binding of operational elements for the case of

      i. Minimum resources: one adder (A1) and one multiplier–divider (MD1)

      ii. Maximum resources: three adders (A1, A2, and A3), two multipliers (M1 and M2), and one divider (D1)

   c. Calculate function execution time for both cases given earlier with minimum and maximum resources and provide respective timing schedules.

   Note: The adder requires one clock cycle for operation and the multiplier or divider needs four clock cycles for respective operations.

Project

Specification

The *luminance component* Y of $Y'CbCr$ in RGB encoding process is one of the required functions in the MPEG video-transmission systems. The Y_i value is encoded for every pixel of the video frame. For digital (8 bits/sample) YCbCr,

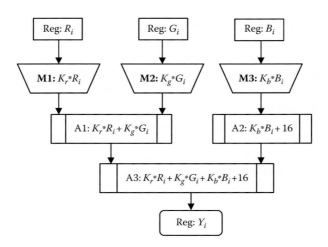

FIGURE 9.P.1
Architecture of the data-path of YCbCr color convertor.

the following (simplified) formula is given for Y_i computation: $Y_i = 16 + K_r * R_i + K_g * G_i + K_b * B_i$.

R_i, G_i, and B_i are values (8-bit) of red, green, and blue components of each pixel in a video-frame where i is the pixel number in 1080 p video-frame format that progressively increases from 1 to 2,073,600 (1920 pixels per row × 1080 rows).

K_r, K_g, and K_b are the constant coefficients (8-bit) determined in video-conversion standard.

The architecture of the YCbCr application-specific processing (ASP) module is presented in Figure 9.P.1. This ASP consists of three multipliers (M1, M2, and M3) and three adders (A1, A2, and A3) as operational elements composing fully pipelined data-path.

Problem 1 *ASP Data-Path Performance*
Determine the latency and cycle time in the case of fully pipelined ASP data-path shown on the block diagram in Figure 9.P.1 when

1. Each multiplier performs multiplication operation in 2 c.c.
2. Each adder performs operation in 1 c.c.

Provide Complete Timing for the Three Cycles of Y Calculation (Y_1, Y_2, and Y_3) and show the timing diagram in Figure 9.P.2 (name of the resources are given according to the block diagram in Figure 9.P.1). The content of the figure cell shows the number of RGB vector under processing (e.g., $i = 1 \rightarrow$ [1]).

Note: The next operation in the pipeline of the same vector can start only after completion of the previous operation (e.g., A1 on RGB [1] can start its operation after completion of M1 and M2 on the data of the same vector).

Resources	Clock Cycle #										
	1	2	3	4	5	6	7	8	9	10	
M1	[1]										
M2	[1]										
M3	[1]										
A1			[1]								
A2			[1]								
A3				[1]							

FIGURE 9.P.2
Timing diagram: schedule for all ASP resources per clock cycles.

Problem 2 *Timing Characteristics of a YCbCr Convertor*

1. Calculate the frame processing time *in clock cycles* for all 2,073,600 pixels of the video frame.
2. Calculate the operating clock frequency (in MHz) for ASP that will ensure 120 fps frame rate.

Problem 3 *Input and Output Bandwidth Characteristics*

1. Determine the ASP *input bandwidth,* assuming that
   a. 3 bytes of R_i, G_i, and B_i values must be delivered within one cycle time of Y_{i-1} computation
   b. Coefficients K_r, K_g, and K_b are hardwired to the ASP multipliers and thus there is no need to input their values per frame
   c. The input bandwidth (bytes/second) is equal to the number of bytes to be transmitted to ASP per second

   Note: 1 MB = 1024 × 1024 bytes
2. Determine the number of input lines for RGB transmission *to the* ASP circuit

   Number of input lines

   $$= \frac{\text{Number of input bits to be delivered to ASP per cycle time}}{\text{Number of clock cycles in cycle time}}$$

3. Determine the ASP *output bandwidth* taking in account the following:
   a. Result of summation of two binary digits must have one additional (most significant) bit to accommodate possible carry bit resulting the summation (e.g., summation of two 8-bit integers may result a 9-bit word).

b. Multiplication of two digits adds the number of bits of the multiplier to the number of bits of the multiplicand (e.g., 8-bit multiplicand multiplied on 8-bit multiplier results 16-bit word).

c. The output bandwidth (bytes/second) is equal to the number of bytes to be transmitted from ASP per second.

Note: 1 MB = 1024 × 1024 bytes

4. Determine number of output lines for Y_i data transmission *from the ASP circuit*:

Number of output lines

$$= \frac{\text{Number of output bits to be transferred from ASP} \times \text{per cycle time}}{\text{Number of clock cycles in cycle time}}$$

Problem 4 *Creation of Component Symbol*

In the box in Figure 9.P.3, draw the *symbol* for the aforementioned ASP. The symbol should reflect all the input/output buses, control and synchronization signals, clock, and reset signals.

Note 1: For all buses please indicate the number of data lines determined according to the input/output bandwidth required by Y_i execution process cycle (e.g., input 16). For all signals indicate the name and direction (e.g., "Busy" output).

Note 2: Use V_{sync} (vertical synchronization) and H_{sync} (horizontal synchronization) signals as initiation and termination signals for video-frame execution process.

Video clock (VCLK) can be used as the strobe (synchronization signal) for R, G, and B input values. Y output should generate its own synchronization signal (strobe) Y_sync.

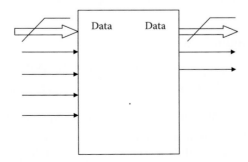

FIGURE 9.P.3
Symbol of the ASP unit dedicated for computing the YCbCr color conversion.

Note 3: YCbCr ASP must inform the rest of the system when it is busy (in Y_i-processing cycle).

Note 4: Reset signal must be active low and asynchronous as global termination signal for all processes in ASP circuit.

Note 5: Global clock signal should provide operating frequency to all circuits of ASP.

References

1. R. G. Bartle, *The Elements of Real Analysis*, John Wiley & Sons, New York, 1967, pp. 11–13.
2. D. C. Mead, *Direct Broadcast Satellite Communications: An MPEG Enabled Service*, Addison-Wesley Wireless Communication Series, Prentice-Hall, Inc., Upper Saddle River, NJ, 2000, Part 3, Chapters 7 and 8, pp. 115–170.
3. J. Davidson, *An Introduction to TCP/IP*, Springer, New York, December 1989, 112pp.
4. D. C. Mead, *Direct Broadcast Satellite Communications: An MPEG Enabled Service*, Addison-Wesley Wireless Communication Series, Prentice-Hall, Inc., Upper Saddle River, NJ, 2000, Part 5, Chapter 13, pp. 229–240.
5. E. P. J. Tozer (Ed.), *Broadcast Engineer's Reference Book*, Focal Press (imprint of the Taylor & Francis Group), Burlington, MA, 2013, Chapter 2.8, pp. 233–264.
6. J. L. Hennessy, D. A. Patterson, *Computer Architecture: A Quantitative Approach*, 3rd ed., Morgan Kaufmann Publishers, San Francisco, NC, 2003, Chapter 3 and Appendix-A.
7. Xilinx Inc., *Virtex-6 FPGA Configuration*, User guide, UG360 (v 3.7), Xilinx Inc., San Jose, CA, November 2013, p. 87, Table 6-5.
8. Xilinx Inc., *LogiCORE IP MicroBlaze Micro Controller System*, Product guide, PG048 (v1.4), Xilinx Inc., San Jose, CA, March 20, 2013.
9. ALTERA Corp., Nios II Hardware Development Tutorial, TU-N2HWDV-4.0, Altera Corp., San Jose, CA, May 2011.
10. G. De Micheli, *Synthesis and Optimization of Digital Circuits*, McGraw-Hill, Inc., New York, 1994, Chapter 3, pp. 97–138.
11. G. De Micheli, *Synthesis and Optimization of Digital Circuits*, McGraw-Hill, Inc., New York, 1994, Chapter 4, pp. 139–184.
12. M. M. Mano and C. R. Kime, *Logic and Computer Design Fundamentals*, 2nd ed., Prentice-Hall, Inc., Upper Saddle River, NJ, 2000, Chapters 3–8, pp. 91–466.
13. S. Brown and Z. Vranesic, *Fundamentals of Digital Logic with VHDL Design*, 2nd ed., McGraw-Hill, New York, 2005.
14. Xilinx Inc., *ISE In-Debt Tutorial*, User guide, UG695 (v14.1), Xilinx Inc., San Jose, CA, April 24, 2012.
15. ALTERA Corp., Introduction to the Quartus II Software, Version 10, Copyright 2010, MNL-01055-1.0.

10

Virtualization of Reconfigurable Computing System Architecture

10.1 Introduction

After the discussion of principles of task segmentation and virtualization of function-processing units in the previous chapter, this chapter will describe the principles and methods for the integration of segment-specific virtual components and the complete application-specific processing circuits in spatial and temporal domains.

In contrast to conventional programmable computers with fixed architecture, reconfigurable computing systems can change the set and functionality of system components as well as links between components per time. In other words, system architecture can change morphology (form of components) in the spatial domain and the topology of interconnections between components in the temporal domain in response to the variable requirements of the application and/or conditions in the environment of the RCS. The components can be software, hardware, or hybrid in nature depending on the performance requirements of a task or the external conditions influencing the system. Thus, flexibility in system architecture can provide runtime adaptation to variations of the workload and/or external conditions. This, in turn, can significantly increase cost efficiency, power economy, reliability, and sustainability of the computing systems based on the principles of RCS. However, the aforementioned dynamic nature of the RCS architecture dictates the necessity of specific mechanisms for the static and dynamic integration of components into the processing architecture. This chapter also considers the mechanisms of architecture integration. There are many different possible variants of RCS architectures as well as methods of architecture integration. There are many literature sources describing the different ways of RCS architecture organizations, their pros and cons, as well as application orientation (e.g., [1,2]). It would be difficult to observe all the existing reconfigurable architectures and methods for the static and dynamic integration of RCS architectures. Thus, this chapter focuses on the main concepts of architecture organizations that already have or may have practical applicability.

10.2 General Organization of the RCS Architecture

Since an application in general case may have algorithmically intensive and computationally intensive segments, the RCS must contain a set of associated virtual components:

1. Virtual software components (VSCs) to be processed on the respective ALU-based platform (e.g., CISC or RISC processors) for algorithmically intensive task segments
2. Virtual hardware components (VHCs) to be implemented in the field of configurable resources (FCR) as function-specific processing units (FPUs) for computationally intensive task segments
3. Virtual hybrid components (VHbC) to be implemented as a combination of VSCs running on CISC or RISC processor tightly integrated with dedicated hardware accelerator(s) in the form of application-specific processors (ASPs) configured in the appropriate PLD

Therefore, in general, the RCS architecture should consist of two conceptually different parts:

1. The part oriented on processing of algorithmically intensive segments of a task
2. The part oriented on processing of computationally intensive segments of a task

Each part must contain appropriate data-processing subsystem, memory, and I/O subsystems, and cooperate with another part over the communication infrastructure (e.g., bus or network). The general organization of the RCS architecture is presented in Figure 10.1.

This architecture can be divided into three parts: (1) the data-processing system, (2) system memory hierarchy, and (3) I/O interface subsystem. Let us consider each of the previously listed parts:

1. *The data-processing system* consists of two parts:
   a. The part based on *instruction processor(s)* for algorithmically intensive functional segments presented in a form of software routines (VSC files). This part may be implemented as single-core or multicore sequential processors, each of which may have CISC- or RISC-type architecture. These processing cores can be implemented in three forms: (i) stand-alone fixed components (e.g., microprocessor chips), (ii) on-chip hard cores as built-in processors with fixed architecture, and/or (iii) on-chip soft-core processors configured inside the PLD in one of the dedicated places for some period of time. The processing cores are

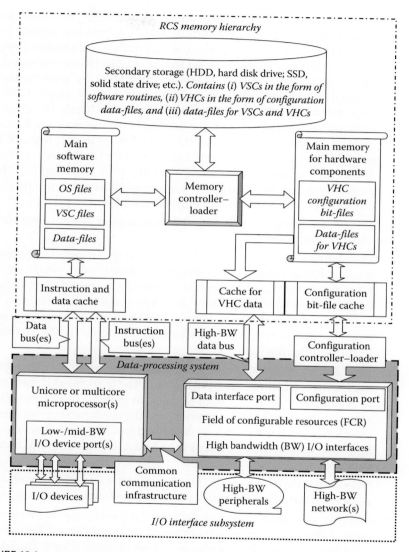

FIGURE 10.1
General organization of system architecture of RCS.

connected to the memory hierarchy via instruction and data buses. Also, the aforementioned processors usually have dedicated low-bandwidth and/or mid-bandwidth interfaces to different I/O devices (e.g., UART, I2C, SPI, and GPIO).

b. The part based on *configurable logic resources* (FCR) oriented on computationally intensive functional segments presented in a form of configuration bit-files (VHC-files). This part is organized according to the principles and configuration mechanisms described in Chapters 3 through 8. The FCR here is assumed in all levels from the on-chip level of computing, communication, and memory resources to the system level of the same resources. The FPUs are configured at the on-chip level by loading the respective VHCs to the configuration memory of target PLDs over the configuration port(s). The FPUs can exchange the information to be processed with the external data memory via high-bandwidth data bus(es) using standard bus protocols (e.g., PCI or PCI express) or custom (application-specific) data bus(es). The FPUs can exchange information directly with high-bandwidth peripherals (e.g., video sensors, high-resolution video monitors, high-precision actuators) and high-bandwidth network(s).

2. *System memory hierarchy* reflecting data-processing system specifics consists of three parts:

   a. The *software memory hierarchy* supporting instruction-based processor(s) of the RCS. This hierarchy consists of the main software memory, instruction and data caches, and secondary storage for software routines (e.g., hard disk drives, solid-state drives).

      This part of memory is similar to the memory hierarchy of modern instruction-based processing systems with virtual memory organization.

   b. The *Memory hierarchy for FCR* consists of the main memory for the hardware components (VHCs) and the associated data-files or memory for temporal data storage. In general case, this subsystem may contain cache memory for the configuration bit-files (on-chip or external cache), as described in Chapter 9. Also, configurable data cache or buffers may be added for temporal data storage.

   c. *Secondary storage* containing libraries of VHCs and VSC as well as associated data-files for applications to be executed on the RCS.

3. *Input/output subsystem* that also reflects the previously mentioned division of the RCS on two parts oriented on algorithmically intensive and computationally intensive segments of applications. The I/O interfaces oriented on instruction-based processors are usually standard communication interfaces such as serial USB, SPI, or network (e.g., Ethernet) interfaces. In contrast, the I/O interfaces providing data exchange for the FCR are usually customized to the application of high-bandwidth communication channels. These channels can serve as interfaces for high-bandwidth peripherals: high-resolution/high frame rate video sensors, HD video monitors, high-precision actuators, etc. Also, these interfaces can work as communication channels for multigigabit networks.

In other words, the RCS architecture in its general form looks like a combination of modern general-purpose instruction-based computing systems and a set of PLDs. This FCR-based part allows effective utilization of hardware resources for tasks where large volumes of data must be processed in a relatively short period of time.

Since the architecture of software processing is similar to the architectures of instruction-based CISC or RISC processors, the following sections will describe the main types of RCS architecture focusing on the FCR-based part and the main architecture organizations of processing circuits implementing VHCs.

10.3 Concept of ASVP and Hardware Components Integration

The main problem for the synthesis of the processing circuits optimized for the application is the integration of hardware and software components to the complete architecture of the ASP.

Definition 10.1 *Application-specific processor (ASP)* is the data-processing circuit optimized to the specifics of the application's algorithm and data structure satisfying all performance specifications and system constraints for a given application (task).

ASP is a *hardware circuit* that can be implemented in the form of

1. ASIC—application-specific integrated circuit or a set of ASICs (chip set)
2. Circuit configured in the PLD(s)
3. Circuit combined by ASIC (hard core) and configured circuit in PLD (soft core)

The concept of RCS considers ASP as the temporal hardware unit. In other words, the architecture of ASP exists only for a period of time while the application needs it. ASP consists of FPUs required for those task segments that should be performed within a given mode of operation and associated functionality. There could be some FPUs common for all modes of operation and thus included in the ASP architecture at all times. The components reflecting these FPUs are called *static*. These components can be implemented as ASIC(s) on a system level or as a *static part of the SoC* configured in the FCR.

On the other hand, there could be components that may be needed only for certain modes of operation and therefore may not be included to the system architecture at all times. These components are called *dynamic components,* and their configuration bit-files may be stored in memory until request for activation. The dynamic components could be as follows:

1. Hardware components represented as VHCs in the form of configuration bit-files
2. Software components represented as VSCs in the form of software routine files

The concept of RCS assumes that not all hardware components are present in the system architecture at all times. Instead, only static hardware components are configured in RCS from its initialization up to the end of task processing. The dynamic components being represented as VHCs are stored in configuration memory in virtual form (configuration bit-files). Hence, this type of ASP can be defined as the application-specific virtual processor (ASVP).

Definition 10.2 The *ASVP* is the data-processing architecture that consists of a static part allocated in actual hardware circuits and a dynamic part allocated in memory as VHCs in the form of configuration bit-files.

Thus, the hardware resources of FCR in RCS are used only for the static part of the ASVP and the dynamic components configured in FCR only for the active mode of operation. This approach enables adaptation to variations of (1) workload, (2) performance requirements, and (3) external environmental conditions with minimum hardware resources and power consumption.

In other words, in ASVP-only static components are deployed in hardware at all times. The rest of the components are loadable to the configuration memory of the FCR according to the requirements of the mode of operation and/or system constraints. Hence, the hardware architecture of ASVP is not static by definition but varies in time. Furthermore, in contrast to ASICs and statically configured ASPs, there are two processes in ASVP assembled in parallel or one after the other: (1) process of data execution and (2) process of architecture reconfiguration.

The dynamic nature of the ASVP architecture dictates the necessity of hardware component integration into the mode-specific ASP architecture. This process for multimodal applications can be done (1) statically (prior to data execution) or (2) dynamically (in runtime of data execution).

The static integration of components assumes compilation of the ASP architecture for each mode of operation offline. In this case, the integration of static and dynamic components (FPUs) for each of operational mode must be performed prior to data execution. Therefore, for each mode of operation, it is necessary to

1. *Determine all task segments* to be executed in a given mode and thus their associated functions
2. *Create a mode-specific ASP* architecture according to the set of FPUs required for the mode
3. *Compile the mode-specific ASP* as the mode-optimized variant of ASVP and store it in the form of a configuration bit-file to be loaded to the FCR when the mode is activated

Therefore, static component integration assumes offline ASP architecture compilation in a manner similar to the compilation of ASIC-like computing systems deployed in PLD(s). Generally speaking, this process is similar to the architecture synthesis and logic design of the set of ASPs to be allocated on the same PLD platform. This approach is effective when (1) The number of modes is relatively small and (2) The required mode-switching time is equal or greater than the reconfiguration period of FCR for this mode.

The main advantage of component integration into the mode-optimized ASPs is the relatively simple architecture synthesis and offline compilation process. This method allows preexecution verification of system functionality and performance characteristics in different environmental conditions (e.g., temperature or power level variations). Another advantage is the relatively simple implementation of this ASVP on the PLD-based platform.

The main drawback of this approach is the necessity to stall the system completely during the period of reconfiguration to the new mode and the relatively long mode-switching time. Another limitation for ASVP implementation is the number of possible modes (e.g., 50+). In many practical cases, the aforementioned aspects are the main obstacles for the utilization of system integration. Nonetheless, the multimodal RCS based on the principles of static integration of mode-specific ASPs is a popular architecture solution. The organization of this type of RCS will be discussed in detail in Section 10.4.

An alternative to the static offline integration of functional components to the mode-specific ASP architecture is the dynamic (runtime) integration of the ASP architecture.

Definition 10.3 *Runtime ASP integration* is the process of composition of mode-specific processing circuit in the part of FCR while the rest of the circuits in the FCR can continue noninterrupted data-processing.

Dynamic integration of mode-specific ASP can be the most effective solution in cases when

1. The number of operational modes is in the hundred to thousand range.
2. Mode-switching time is less than the reconfiguration time of the complete FCR.
3. The on-going data-execution processes of the application cannot be interrupted.

Obviously, the static component integration and generation of thousands of mode-specific configuration bit-files in such cases can be unacceptably time consuming and expensive. In Section 2.5, it was shown that the utilization of resources in dynamically partially reconfigurable FCR is higher than in statically configurable systems. Furthermore, the resource utilization factor increases proportionally to the number of possible modes. Considering the mode of operation, according to Definition 2.12 in Section 2.4 as one of the possible combinations of the task algorithm, data structure, and set of constraints, the number of operational modes in many practical cases can be relatively high.

Let us consider the example of application where (1) the number of variants of the task algorithm is 16 (e.g., video-processing modes), (2) the types of data structure (e.g., video resolution and number of bits per pixel) for each mode is 8, and (3) a set of performance constraints (e.g., possible frame rates) is 4. The number of possible modes will be equal to $16 \times 8 \times 4 = 512$. Hence, if static integration of FPUs is used to implement this application, 512 mode-specific ASPs must be compiled, resulting in 512 configuration bit-files composing the ASVP.

When operational temperature, level of power, etc. is taken into account, the number of modes can reach many to thousands. This can become a significant barrier for utilization of the approach based on static component integration.

Also, the response time associated with the mode-switching period can be much shorter in the case of dynamic partial reconfiguration because the volume of partial configuration bit-files is much smaller than the complete PLD configuration bit-file. The aspect of configuration time for partial reconfiguration was analyzed in Chapter 7 stating that smaller configuration bit-file provides shorter reconfiguration time if the same configuration port and bus is used.

The only drawback of this approach is the relative complexity of the run-time ASP integration process. Indeed, the dynamic integration of FPUs into the complete ASP architecture not only assumes the configuration of FPUs in the FCR by loading respective VHCs to the configuration memory of FCR but also requires several processes to be performed in the period of integration:

1. *Scheduling the FPU* configuration process in predetermined location of the target PLD
2. *Changing the topology* of component interconnection links (or buses) to provide the recently loaded FPU with up-stream (input) data and transmission of the down-stream data (results) from this FPU to the rest of the system

3. *Synchronization of the FPU's* I/O interface with other components avoiding potential multisourcing or other structural hazards in the data-exchange process

4. *Determination of proper moments of time for FPU initiation and termination* avoiding the violation of the entire computation process in the RCS

There are several methods allowing safe organization of the run-time integration of FPUs to the ASP architecture each of which determines the class of the run-time reconfigurable (RTR) RCS. Main methods of the dynamic integration of ASP in the FCR are the following:

1. *FPU to CPU integration* over the common bus. This method is usual for hybrid RCS often called—*application-specific instruction processors (ASIP)*. The ASIP can contain a set of hardware accelerators for computationally intensive macro-functions. These accelerators can be implemented in FPUs to be configured in the dedicated areas of the target PLD. These FPUs can be stored in respective VHCs in the external configuration bit-file memory and activated by special instructions executed in the CPU. These instructions can initiate (a) loading the required VHC's configuration bit-file to the reserved area(s) of the PLD; (b) interfacing the VHC-based FPU (hardware accelerator) to the CPU, data memory, and/or I/O interfaces; and (c) data-execution and data-transfer processes in the requested FPU according to the associated macro-function. Integration of these FPUs to the ASP looks like the *plug-and-play* process for configured FPU(s) and CPU-centric architectures built around a common bus or backplane. This type of systems will be discussed in Section 10.5.

2. *FPU to data memory integration* over the dedicated memory bus. This method is usual for RCS with *temporally partitioned resources* where several FPUs are configured in the FCR one after another (sequentially) and store their temporal results in the associated memory unit(s). The integration process here is quite simple because neither dynamic scheduling nor changing of the interconnect topology is needed. Data transactions here are always dedicated between the FPU and respective memory module at a time. Therefore, structural and timing conflicts are excluded by default as well as multisourcing. This type of RCS will be described in Section 10.6.

3. *Multi-FPU integration* over the on-chip communication infrastructure. This method assumes the most flexible organization of the ASP architecture, where FPUs can be aggregated according to the data flow and control dependencies. This method also allows reaching the highest data-execution rate and reliability due to run-time restoration from hardware faults. This type of RCS will be presented in Section 10.7.

10.4 ASVP with Statically Integrated Hardware Components

One of the most popular solutions for multimodal RCS design is statically reconfigurable computing systems. This type of RCS is based on statically configurable FCR for each mode of operation discussed in Section 2.4. From an application point of view, the multimodal task, T, can be divided into multiple subtasks $T = \{T_i\}$ where each subtask—T_i, is associated with a particular operational mode, M_i. For each subtask, the mode-specific ASP architecture can be designed and compiled: $T_i \rightarrow ASP_i$, $i = 1, 2,...,n$. The ASVP for the task consists of configuration bit-files of the mode-specific ASPs and the mode-selection unit. The ASVP can be implemented in statically reconfigurable RCS as was discussed in Section 6.3. The configuration bit-files of all ASPs are stored in the external memory for configuration bit-files. The mode-selection segment of the task can be deployed in the controller–loader as shown in Figure 10.2.

Let us consider ASVP implementation on the statically reconfigurable RCS platform using the example of application discussed in Chapter 9. The task algorithm was shown in Figure 9.1.

Two modes of operation can be determined in this task (depicted in Figure 10.3):

1. Mode-1 that requires execution of functions F_2 and F_4 or F_m
2. Mode-2 that needs execution of functions F_3 and F_i or F_m

The mode-selection segment of the task consists of the initial routine and conditions check function—F_1 and associated mode-condition statements: (1) C_{init_2} for Mode 1 initiation and (2) C_{init_3} for Mode 2 initiation.

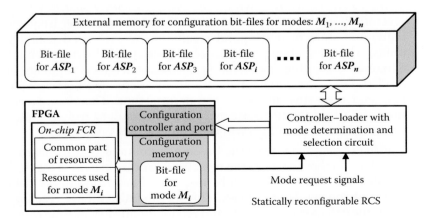

FIGURE 10.2
ASVP implementation on the platform of statically reconfigurable RCS.

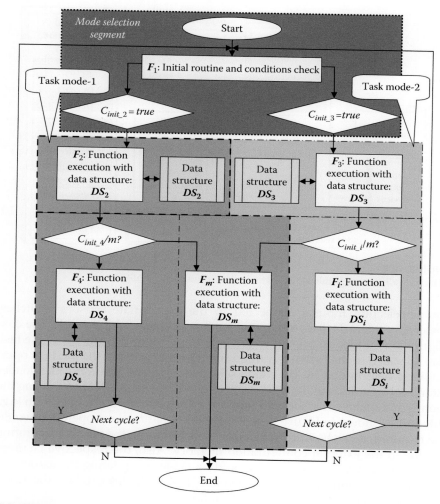

FIGURE 10.3
Task segmentation for the mode-selection segment and Mode 1 and Mode 2 segments.

In case of ASVP implementation on the statically reconfigurable RCS, the mode-selection segment is implemented as static mode-determination and selection circuit and should run at all times on the system. The RCS architecture combines (1) the controller–loader with the mode-selection circuit responsible for FCR reconfiguration and (2) the PLD-based FCR for data-processing according to the mode of operation. The block diagram of this architecture is shown in Figure 10.4.

The mode-selection segment of this task can be deployed in the controller–loader in the form of an F_1-specific FPU. The controller–loader can receive information necessary for the determination of the mode of operation via

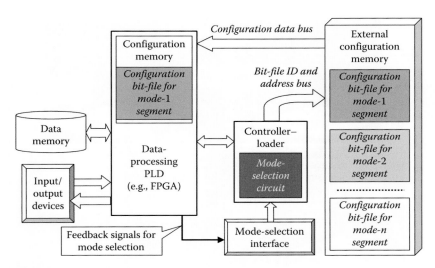

FIGURE 10.4
The statically reconfigurable RCS accommodating multimodal ASVP.

mode-selection input(s) connected to the appropriate sensors, switches, external control devices, or internal feedback signals from the PLD. The configuration bit-files associated with Mode 1 and Mode 2 segments of the task are allocated in the external configuration memory. When the mode of operation is determined by the mode-selection segment, the controller–loader starts configuration of the processing circuit in the target PLD by loading the configuration bit-file associated with the selected mode of the operation (e.g., Mode-1 as shown in Figure 10.4). If any additional mode of operation would be required, it can be simply added by programming the new configuration bit-file to the external configuration memory. Also, the respective changes in the mode-selection segment should be made by updating the associated circuit in the controller–loader. Therefore, the ASVP designed for the aforementioned application consists of the following:

1. Static component (F_1-specific FPU) for mode selection/determination constantly allocated in the controller–loader or similar functional unit and
2. Set of virtual processing components (e.g., VHC_1 for Mode-1 execution and VHC_2 for Mode-2 execution, respectively) stored in the external memory for configuration bit-files.

The process of loading the aforementioned VHCs to the data-processing PLD and further integration with the hardware platform is conducted by the controller–loader according to the mode-selection segment of the application. The controller–loader also performs initiation and termination of data-processing circuits in the PLD according to the mode-selection function.

The initiation, termination commands, and the feedback signals from the data-processing PLD according to the requests for the next cycle are conducted via the control bus between the controller–loader and data-processing PLD.

The architecture of the statically reconfigurable RCS accommodating this multimodal ASVP shown in Figure 10.4 can be generalized to the systems with any number of modes where each mode of operation is implemented in respective VHCs stored in the form of configuration bit-file.

10.5 ASVP with Dynamically Integrated Software and Hardware Components

As was mentioned in Section 10.2, in general, an application may contain algorithmically intensive and computationally intensive segments. In cases when an application contains segments where computation results should be obtained within a period of time shorter than the CPU core can provide, the *hardware acceleration* of those computationally intensive segments is needed. As was shown in Sections 1.4 and 1.5, computing systems with a programmable procedure are limited in their performance by sequential nature of data execution and timing overhead caused by the instruction execution process. Hence, the computationally intensive segments in such applications should be processed by dedicated hardware circuits optimized for the algorithm and data structure of these segments. In other words, if majority of task segments does not need acceleration, they can be programmed as software components (VSCs) to be processed on the instruction-based processor(s). The computationally intensive segments should be implemented as VHCs to be processed on VHC-based FPU(s) deployed in PLD. This class of RCS is considered as *hybrid RCS* where the instruction-based processor is coupled with one or multiple hardware accelerators. The instruction set of this type of hybrid systems include special instructions dispatching computationally intensive segments to the respective hardware accelerators. This class of processors is called *ASIPs* [3]. The hardware accelerators can be implemented as function-specific processing circuits or even macro-function ASPs. The DSP processors that include floating-point multipliers or MAC circuits [4] can also be considered as examples. However, for more general cases where the number of possible macro-functions is high, implementation of those functions in the form of FPUs and circuits constantly deployed in the system can be quite an expensive solution. Therefore, the idea of *reconfigurable ASIP* in the form of an instruction-based processor coupled with reconfigurable hardware accelerators seems promising when the type and number of possible FPUs are not specified up front. In this case, the set of possible FPUs can be presented as VHCs and stored as configuration bit-files in their respective external storage. The general architecture of this class of RCS is presented in Figure 10.5.

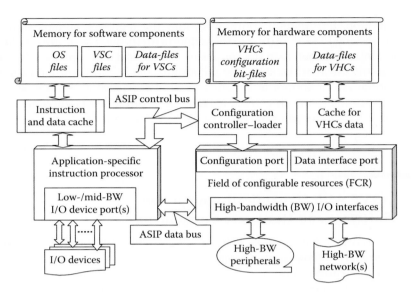

FIGURE 10.5
The RCS based on ASIP coupled with reconfigurable hardware accelerators.

Thus, ASIP-based RCS architecture allows processing of algorithmically intensive segments of the task and computationally intensive segments simultaneously. The task execution process starts from loading initial software components (OS-files and/or VSC-files) from the memory for software components to the ASIP. Then, ASIP starts the program file(s) execution. These program files (VSCs) may contain special instructions or subroutines that can request loading hardware components (VHCs) to the target PLD coupled with the ASIP and interconnected via the ASIP data bus. The selected VHC being loaded configures the respective FPU that provides execution acceleration of the given data structure (e.g., data vector, massive data, or data-stream).

The VHC loading process and FPU configuration can be performed at start-up time in case the type of computationally intensive operations is known up front and data-files are ready for execution. Alternatively, the FPU configuration can be performed ad hoc in cases when the request for function cannot be predicted by application. Since, the FPU configuration requires some period of time, this time should be added to the data-file execution time as overhead. Otherwise, the acceleration of data execution will not be fair in comparison with procedural data execution performed on the instruction-based processor. For example, if procedural execution time of a data-file is—T_{vsc}, and the execution of the same data-file on the dedicated FPU—T_{fpu}, the speedup (acceleration of data execution on the FPU) will be as follows: $R_{vhc} = T_{vsc}/(T_{vhc} + T_{conf})$ where T_{conf} is the FPU configuration time in the target are of the PLD.

To minimize or avoid this overhead associated with FPU configuration, a certain prediction mechanism is needed. This is often possible because each task segment is initiated by a predetermined set of conditions. Thus, when these conditions are expected to be satisfied, the FPU(s) associated with this task segment can be configured.

The concept of this class of RCS was proposed in 199X. One of the first architectures of these types of systems was PRISC [5] where the RISC processor was coupled with programmable FPUs. The GARP [6] is another example of hybrid system where C-language compiler could automatically perform partitioning the task on software segments (VSCs) and segments to be executed on the attached PLD (VHCs). Then, the on-chip implementation continued utilization of this concept of RCS. One of the first system-on-programmable chip (SoPC) utilized the hybrid architecture was NAPA-1000 [7] composed of a 32-bit compact-RISC processing core (CR-32) and adaptive logic processor ALP. The programming could also be done using special C-compiler allowed tagging the task segments to be compiled to the hardware components. To support this class of RCS architecture, Xilinx has included the PowerPC core to all families of Virtex FPGA devices from Virtex-II Pro to Virtex-6. Recently, the Xilinx Zynq FPGA with RISC-based ARM-Cortex core [8] is specifically designed for convenient implementation of hybrid RCS architectures with configurable hardware accelerators. The ARM processing core is also included in several ALTERA FPGA architectures: Altera Cyclone-V and Stratix-10 FPGA families with ARM-based HPS: hard processing system on chip [9].

In other words, the most effective way for processing the "hybrid" applications consisting of algorithmically and computationally intensive parts requires utilization of hybrid processing platform. Therefore, programming of these applications assumes coding:

1. Algorithmically intensive segments in procedural languages to be compiled to the VSCs.

2. Computationally intensive segments in hardware description languages to be compiled to VHCs.

To illustrate the aforementioned, let us consider an example of an application shown in Figure 9.1. If the functions F_1, F_2, and F_3 of this application are algorithmically intensive and functions F_4, F_i, and F_m are computationally intensive, the composition of F_1, F_2, and F_3 functions can be programmed in one software segment, VSC_1, and each of F_4, F_i, F_m function can be compiled to the configuration bit-files of hardware components, VHC_4, VHC_i, and VHC_m, respectively, as shown in Figure 10.6 in gray color.

The software and hardware components of this application are composed of the (ASVP) consisting of VSC_1, VHC_4, VHC_i, and VHC_m.

According to the ASIP-based RCS architecture (Figure 10.5), the VSC_1 is stored in the memory for software components. When initiated, it will run

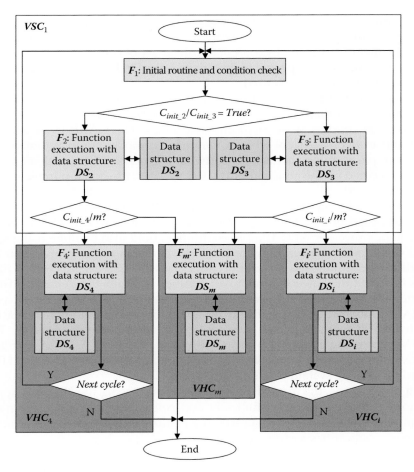

FIGURE 10.6
Partitioning the task on software and hardware components.

on the instruction-based processor (ASIP). This assumes that program files from the VSC_1 will be loaded block-by-block from the memory to the instruction cache of the ASIP to be sequentially executed instruction-by-instruction in the CPU of the ASIP. The data-files required for functions F_2 and F_3 are also stored in the memory for software components in capacity of the data structures DS_2 and DS_3 associated with VSC_1.

When the condition $C_{init\_2} = true$, the function F_2 starts working. Then, if condition C_{init_4}/m selects a computationally intensive function, F_4, the associated FPU must be initiated.

The process of F_4 initiation and further data execution requires dynamic integration of the hardware component VHC_4 with the static part of the RCS according to the steps described in Section 10.3:

1. *Scheduling the FPU_4 configuration process* is initiated by a special instruction in the program segment performing F_2 or special subroutine calling the process of loading VHC_4 bit-file to the target PLD over the configuration port. This instruction or subroutine sends the requested VHC number to the configuration controller over the ASIP control bus (Figure 10.5). Then, the controller–loader starts loading the configuration bit-file of the VHC_4 to the PLD. When done, the controller–loader sends the feedback to the ASIP.

2. *Changing the topology of component interconnection* consists of the establishment the interfaces to the
   a. Peripherals attached to the PLD's I/O ports
   b. Data bus interconnecting the PLD and ASIP data ports.
      Both types of interfaces are included into the VHC design and appear when the configuration of the associated FPU is completed.

3. *Synchronization of the FPU's I/O interface* with other components avoiding potential multisourcing can be done in the usual plug-and-play interfacing. This assumes a certain procedure of initial interaction between a recently configured FPU_4 and data bus of the ASIP. This procedure can be performed by VHC calling subroutine or other processes specifically designed for FPU plug-in. The synchronization of the data bus of the ASIP and FPU takes time, which can be considered as an addition to the configuration time overhead.

4. *Determination of proper moments of time for FPU initiation and termination* avoiding violation of the entire computation process in the RCS. This process can be controlled by the program (VSC_1) and/or by the FPU (FPU_4). In case of CPU-controlled initiation and termination of the FPU operation, special commands must be sent over the data bus to the FPU. In this case, FPU should work in slave mode. Alternatively, the external synchronization signals or events may initiate or terminate/interrupt the FPU's operations. For example, the vertical or horizontal signals synchronizing the video frame can be used as the initiation or termination signals for the video-processing FPU.

Meanwhile, the feedback from the FPU to the ASIP coordinates further data execution on the ASIP's CPU according to the results received from the FPU (FPU_4). When the function (F_4) is not needed anymore, the existing FPU_4 can be replaced by the new FPU. This process should start from the safe de-activation of the FPU_4 in the target area of PLD. For example, when the condition statement C_{init_4}/m selects function F_m, the respective configuration bit-file of the VHC_m will be loaded to replace the existing bit-file of the VHC_4 in the configuration memory of the PLD. Taking into account that this replacement procedure can affect the on-going data-execution process running in the PLD, the ASIP must stop this data execution by sending the appropriate commands to the currently deployed FPU in the PLD. Thus, any FPU must have application-specific initiation and termination circuits and registers to receive, decode, and perform commands from the ASIP that start and stop data execution in the FPU. At the same time, these commands should be able to tri-state all output circuits of the FPU from the data bus to avoid multi-sourcing. In other words, the initiation and termination subroutine running on the ASIP's CPU should be used at any time when a new FPU is requested. The time of this subroutine execution causes the additional time overhead. Thus, the time,—$T_{exe(i)}$, for the actual execution of a data-file on the FPU_i, $i = 1, 2,\ldots, m$, is equal to the sum of the data-file execution, $T_{fpu(i)}$; configuration time for FPU_i—$T_{conf(i)}$; and FPU integration and disconnection time, $T_{intgr(i)}$:

$$T_{exe(i)} = T_{fpu(i)} + T_{conf(i)} + T_{intgr(i)} \tag{10.1}$$

As per the earlier equation, the utilization of the FPU-based accelerators is effective when the data-execution time on the FPU is much greater than the configuration and integration time needed for this FPU. Since all FPUs should be located at the same PLD or reconfigurable region of PLD, this condition is associated with configuration bus and port organization as well as the complexity of the integration, initiation, and termination processes:

$$T_{fpu(i)} \gg T_{conf(i)} + T_{intgr(i)} \tag{10.2}$$

Thus, reduction in configuration time using the parallel configuration bus and port, PLD-based configuration controller–loader, and high-bandwidth memory for hardware components can be an effective solution for this class of RCS.

In case of RSC utilized for stream-processing applications, when the data-file to be processed is almost unlimited, the configuration and integration time do not influence the effectiveness of the system because condition (10.2) is satisfied by default. As an example, it is possible to mention the processing of video streams, streams of network packets, RF signal processing, and generation.

10.6 ASVP with Temporarily Integrated Hardware Components

In Section 10.4, the statically integrated ASVP was considered. The integration of VHCs in this class of ASVPs is made according to the mode of operation prior to the configuration in PLD. Thus, ASVP consists of a set of mode-specific ASPs represented in the form of configuration bit-files and stored in memory for configuration files. When one of the operational modes is activated, the associated configuration bit-file is loaded to the PLD configuring the mode-specific ASP in the FCR or this PLD.

In the meantime, the CPU-centric RCS with reconfigurable hardware accelerators utilizes the concept of dynamic integration of function-specific units with a CPU-based processing core. In turn, it allows significant simplification of the design process due to software/hardware component partitioning and further implementing them in the form of VSC and VHC accordingly. Further, acceleration of data execution for computationally intensive task segments being performed by FPUs acting as hardware accelerators allows reaching the required performance. This approach is effective when a majority of task segments are algorithmically intensive, and only few segments of the task require acceleration of data execution. On the other hand, the area of the PLD reserved for FPUs is relatively small in comparison with the statically integrated mode-specific ASPs that make these architectures cost-efficient for the aforementioned class of tasks.

Nonetheless, there are classes of applications where most of the task segments are computationally intensive and execution on the CPU-processing cores cannot satisfy performance requirements. On the other hand, the implementation of these tasks in the PLD in the form of a monolithic ASIC-like circuit is not feasible due to area, power, or cost constraints. In these cases, the time-multiplexing of PLD resources is needed to accommodate the required functionality and area/time constraints.

For example, let us consider the video-stream processing task based on 720 p × 60 fps standard. According to this standard, video-frame resolution is 1280 × 720 pixels progressively scanned in rate of 60 fps. Let us consider the same task as was discussed in previous Sections 10.4 and 10.5. For the previously mentioned video-processing task, the data structures associated with all functions are based on the video-frame structure according to the above 720 p standard.

The examples of functions $F_1, F_2, ..., F_m$ could be video-frame acquisition, buffering in video memory, color conversion, edge detection, video-compression functions, etc. According to Figure 10.3 presented in Section 10.4, there can be several modes of operation (e.g., video-processing modes) depending on conditions C_{init_2}/C_{init_3}, C_{init_4}/m, C_{init_i}/m, etc.

Let us consider Mode 1 when $C_{init\_2} = true$ and $C_{init\_4} = true$ too. Thus, in this mode, functions F_1, F_2, and F_4 are active and each process completes a 720 p video frame.

For implementation of the aforementioned task, let us consider the PLD that can provide the on-chip clock frequency up to 333 MHz (3 ns per clock cycle).

If the implementation is done in a form of ASVP with statically integrated components (as discussed in Section 10.4), each mode of operation (subtask) should be implemented as mode-specific ASP. Thus, for Mode 1 performing functions F_1, F_2, and F_4, the ASP_1 with pipelined data path should be designed. In the pipelined data path, the cycle time for one pixel data-processing can be reduced down to one clock cycle per pixel (subject of amount of resources arranged for the ASP). Hence, in case of maximum performance, each 720 p video frame can be executed in 1280×720 pixels $\times$ 3 ns/pixel = 2.765 ms. However, the required data rate for the considered task is 60 frames per second. Therefore, each video frame should be executed in 1 s/60 frames = 16.666 ms, which is more than six times longer than the performance that could be provided by the PLD on its clock frequency.

Thus, this type of the task implementation provides much higher performance than what is really needed for an application. Meanwhile, the amount of resources utilized for this implementation is higher than needed because the ASP must accommodate all processing circuits for functions F_1, F_2, and F_4 working concurrently. The alternative solution based on ASIP approach (discussed in Section 10.5) will not be feasible because all functions in the considered task are computationally intensive. Indeed, if the CPU-core should process a video frame in 16.666 ms, each pixel processing time should not exceed 16.666 ms/(1280×720) pixels = 18 ns/pixel.

Taking into account the sequential execution of instructions and the clock cycle period equal to 3 ns, the available time for one pixel processing should not exceed: 18 ns per pixel/3 ns per clock cycle = 6 clock cycles. During this period of six clock cycles, all three functions, F_1, F_2, and F_4, should be completed. This does not seem to be possible for implementation on the CPU core. The hardware accelerators certainly can perform this job. However, CPU core will not be involved in this process at all.

In case of time-multiplexing of the FCR resources, the implementation of the aforementioned video-processing task is possible in the form of ASVP consisting of VHCs, each of which is associated with their respective function: VHC_1 for F_1, VHC_2 for F_2,..., VHC_m for F_m. Then, these VHCs can be configured one after another in the target PLD. This concept based on temporal partitioning of the PLD resources allows allocation of different segments of the task on the same set of PLD resources in temporal domain. This concept was discussed in Section 2.5.

As was shown in Section 2.5, this approach allows significant reduction of the required PLD resources and, therefore, the associated cost and power consumption.

Let us consider the aforementioned example in light of this concept. In case of Mode-1, the VHC_1 will configure the FPU_1 for initial processing of the video frame according to function F_1 (e.g., video-frame acquisition and buffering). The temporal results of this stage of mode processing should be stored in temporal data memory. When this stage is completed, resources allocated for this task should be reconfigured to the FPU_2 by loading VHC_2 configuration data-file to the same area of the PLD. After that, FPU_2 performs execution of the results of the previous stage of video processing. FPU_2 reads those results from temporal data memory and stores the results in other memory banks of the same temporal data memory. Then, a system again reconfigures

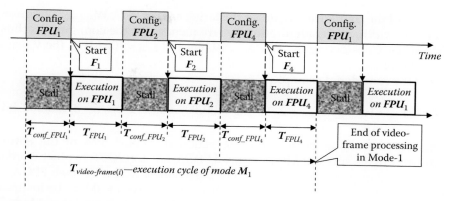

FIGURE 10.7
Execution of video frame on the ASVP with temporarily integrated FPUs.

to the last FPU_4 by loading the configuration bit-file of VHC_4 to the configuration memory of the PLD and video-data execution continues until the end. If the *Next cycle* condition is still required, the aforementioned cycle repeats. The timing of this process is shown in Figure 10.7.

The analysis of performance can be made for the worst-case scenario when each FPU executes all pixels of the video frame. In case of pipelined data path, each FPU will execute the function, within the period of time equal to (1280×720) pixels $\times$ 3 ns/pixel = 2.765 ms.

It is assumed that latency associated with the pipelined nature of FPU data paths is negligible compared to complete pixel processing time. In most practical cases, this assumption is close to reality. For example, if the pipelined data path consists of 100 stages of processing, each of which can be executed in one clock cycle, the total latency of the pipeline will be equal to

$$\text{Latency}(FPU_i) = 100 \text{ stages} \times 3 \text{ ns/clock cycle} = 300 \text{ ns}$$

This is around 0.01% of the total frame execution time on an FPU.

Meanwhile, the total video-frame processing time will not exceed the sum of execution times of all FPUs. In our case, $T_{FPUi} = 1280 \times 720 \times 3$ ns = 2.765 ms. And the total time spent for data-processing of the video frame: $T_{video-frame(i)} = 2.765$ ms $\times$ 3 = 8.295 ms when the required frame execution time for 60 fps frame rate is equal to 16.666 ms.

Hence, the difference between the required 16.666 and 8.295 ms equal to 8.371 ms can be used for FPU reconfiguration. Since there is a sequence of three FPUs, the average reconfiguration time for an FPU—T_{conf}, will be equal to $T_{conf} = 8.371$ ms/3 = 2.790 ms.

It is possible to estimate the volume of configuration bit-file for the respective FPU using Equation 8.1: $T_{conf} = (V_{pld}/(W_{conf\_bus} \times f_{conf}))$, where V_{pld} is the volume of the configuration file for the target area of PLD; W_{conf_bus} configuration bus width, and configuration bus frequency; f_{conf} the configuration bus frequency. Considering the W_{conf_bus} = 32 bits and f_{conf} = 100 MHz, the V_{pld} can be determined.

For the considered example,

$$V_{pld} = T_{conf} \times W_{conf\_bus} \times f_{conf} = 2.790 \text{ ms} \times 100,000,000 \times 32 \text{ bit/s} = 8,928,000 \text{ bit}$$

This volume allows partial reconfiguration of almost 35% of the entire resources in the smallest Xilinx Virtex-6 FPGA—XC6VLX75T. In other words, if one third of processing resources of the smallest Xilinx Virtex-6 FPGA is reserved for this task, the required timing for the given task can be insured in any mode. Indeed, if the condition C_{init_3} and C_{init_i} are "true," then Mode-2 (according to Figure 10.3) is activated. Thus, for the next video frame instead of VHC_2 and VHC_4, the configuration bit-files of VHC_3 and VHC_i will be loaded in the configuration memory of the PLD. Now, the sequence of FPU_1, FPU_3, and FPU_i will perform video-data execution according to functions F_1, F_3, and F_i. Since reconfiguration of the PLD resources is conducted during the period of video-frame processing, mode switching is "seamless" for application. In other words, when conditions are changing, the next video frame can be executed by an updated algorithm. The generic block diagram of the RCS with temporally partitioned PLD resources is depicted in Figure 10.8.

The generic organization of RCS with temporally partitioned (time-multiplexed) resources assumes that all VHCs associated with computationally intensive functions are stored in the external configuration memory in the form of configuration bit-files. The controller–loader and/or mode-selection unit can determine conditions for initiation of one of the FPUs according to the conditional statements in the task algorithm and associated signals

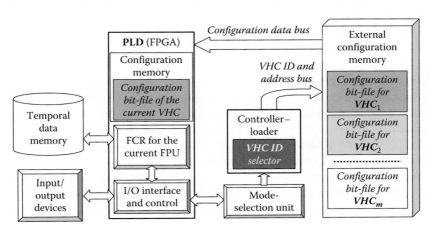

FIGURE 10.8
Generic organization of RCS with temporally partitioned PLD resources.

or flags. Then, the controller–loader sets the ID-code of the FPU that should be initiated. This ID code can be encoded in the most significant address bits of the configuration bit-file address of the correspondent VHC. Since the reconfigurable region in the PLD has predetermined dimensions, all configuration bit-files of VHCs must have an identical volume. Hence, after setting the ID-code, the controller–loader generates the same sequence of addresses to load the configuration bit-file to the configuration memory of PLD.

For an example, considered earlier, if the number of functions is 8 (F_1, F_2,...,F_8), the function ID-code will contain three most significant bits. And if the volume of the configuration bit-file does not exceed 8 Mbit organized as 256 K × 32 bit files, the address area for each VHC: VHC_1, VHC_2,..., VHC_8, will be 18 bits. Thus, the address area in the external configuration memory should have 21 bits: three most significant bits associated with the VHC identification code and 18 bits to address each of, 32-bit word of the VHC configuration bit-file. Therefore, in case of the 32-bit configuration data bus and port working in the 100 MHz bus clock frequency (10 ns per clock cycle) reconfiguration of the PLD resources for the new FPU_i can be done in T_{conf_FPUi} = 256 × 1024 (32-bit words) × 10 ns= 2621 ms. This is less than the available reconfiguration period for the FPU equal to 2.790 ms.

When the configuration file for the VHC is loaded in the configuration memory of the PLD, the associated FPU is configured in the reserved region of the PLD. However, it does not mean that the system is ready to perform data-processing on the FPU. The FPU should be integrated with the rest of circuits according to integration process discussed in Section 10.3. Nonetheless, the integration process in this class of RCS is quite straightforward. Indeed, there is only one FPU as the data-processing component at any moment of time. And for the period of data-processing, the current FPU must communicate only with data memory and/or I/O devices. Therefore, the appropriate interfaces must be included into any FPU design. Since VHC configuration bit-file is equal in volume for any FPU, configuration time is predetermined. Thus, scheduling of the FPU configuration is relatively simple task. The allocation of the FPU in the FCR also is known because only one region is reserved for an FPU. The conflict in memory data bus also is not possible because the memory for temporal data/results always is a slave device working under the control of FPU circuits. Initiation and termination of FPU data execution can be done by an internal (on-chip) control unit or external controller (e.g., part of controller–loader) depending on application specifics. Therefore, it is possible to see that all hardware components (FPUs) being loaded to the PLD are integrated to the rest of the circuits (data memory, I/O devices, control unit, etc.) identically and according to the functional specifics of the FPU. This temporal integration continues for the period between initiation and termination of the FPU. Then, the current FPU can be replaced by the next component. In other words, the ASVP in this class of RCS consists of the (1) static part which includes interface circuits for I/O devices and temporal data and (2) dynamic part which is represented by a set of respective VHCs.

In general, any application consisting of multiple computationally intensive functions can benefit from the aforementioned mechanism of time-multiplexed or temporally partitioned PLD resources. The main benefit is that the application that cannot fit in the available resources in the given PLD can anyway be processed on the aforementioned RCS only for a longer period of time. This aspect was very important at that time when the amount of reconfigurable resources in the PLD was limited (e.g., PACT XPP [10]). Nowadays, this method can also be used in partially reconfigurable FPGA devices for better cost-efficiency of the task execution because it enables minimization of system dimension and cost, static, and dynamic power consumption.

Another advantage of this approach is relative simplicity of task implementation. In case of predesigned RCS platform and the library of macro-functions, programming the task can be simplified to the level when only encoding the VHC IDs according to the task algorithm is needed (e.g., [11]). Any custom functions or macro-functions can be designed using the template provided by the RCS platform vendor. This template should contain the source code (in HDL) of interfaces to the data memory and the possible I/O data ports. There is quite a large volume of academic publications regarding this class of RCS.

In addition to the aforementioned, this class of RCS allows self-restoration of the data path from transient hardware faults. This kind of faults is caused by single-event radiation effects (e.g., single-event upsets, single-event functional interrupts, etc. [12]). The reason of the transient hardware faults is sensitivity of MOS transistors to high-energy subatomic particles. These particles (e.g., protons) can charge gates of CMOS transistors and therefore, change the content of memory cells, logic gates, etc. Since the configuration memory of PLD can be affected by these effects too, the on-chip interconnectivity between gates as well as their functionality (due to the changed content of LUTs) can be jeopardized in run-time. The transient faults can be mitigated by recharging the affected gates or memory cells to restore the proper states. Fortunately, in case of dynamically reconfigurable FPUs, the transient fault can affect performance of only one FPU at a time because the next reconfiguration of the PLD resources to the new FPU will clear up the transient fault automatically. Therefore, no special built-in self-test (BIST) and self-restoration methods are needed for the data path of the RCS with temporally integrated components. There are, however, many methods for the diagnosis and mitigation of the aforementioned class of hardware faults developed for different classes of RCS (e.g., [12,13]).

The main drawback of this approach is the relatively low resource utilization factor RUF. As was mentioned in Section 2.5 for the class of RCS with temporal partitioning of resources on task modes, the RUF can be reduced if the reconfiguration time is close to the data execution time.

In Figure 10.7, the video-frame processing cycle consists of total data execution time of FPUs and total time spent for PLD reconfiguration. Therefore, the ratio between data-processing time and reconfiguration periods (when data execution is stalled) determines the cost-effectiveness of this class of RCS. In other words, reconfiguration time should be considered not only as timing overhead but also as hardware overhead too. Thus, when *the reconfiguration time is close or equal to the total data-execution time*, the *tandem* organization of the RCS architecture can be considered as more effective. In this case, the two separate reconfigurable regions (in partially reconfigurable PLD) or two PLDs are reserved for FPU configuration. During data execution, one PLD or region of PLD accommodates the active FPU when the other PLD or PLD region is in the process of reconfiguration. This method excludes the timing overhead for FPU reconfiguration leaving, however, hardware overhead equal to 50% of reserved hardware resources. Thus, this RCS organization is effective when granularity of functional components and/or volume of data-in data structure are adequate to the volume of configuration bit-file of respected VHCs and bandwidth of configuration bus and port(s) of target PLD(s).

This method can be efficiently used for data-stream processing application such as: video and audio processing, digital signal processing (DSP), network packet processing, and digital video broadcasting (DVB). A specific stream-processing application is the cyclic repetition of identically structured data elements. For example, in video-stream tasks, the data elements associated with pixel (color) brightness are composed of video frames according to certain video standards. In network stream processing tasks, data elements associated with symbols are composed of packets organized in accordance with network standards, etc. The flows of video-frames or network packets are cyclic in nature thus, coming to processor from associated ports one-after-another with associated synchronization signals, in all cases, RCS repeats (1) acquisition of data and storage in memory in a given format, (2) synchronization by signals generated in a data source, (3) composition of output data frames, and (4) transmission according to a requested protocol, etc. These operations should be performed in time range of data-flow and cannot be interrupted by other processes. Thus, system should have the static part with a fixed architecture to provide these operations. Meanwhile, the processing algorithm can change depending on the mode of operation. Hence, this RCS should have reconfigurable part with resources cyclically partitioned on various functional units. This, dynamically reconfigurable part can be organized in the form of *tandem* when timing overhead should be excluded due to performance constraints. An example of this RCS for video-stream processing tasks is shown in Figure 10.9.

This example assumes the on-chip level of *tandem* implementation on partially reconfigurable PLD, where resources can be time-multiplexed on different FPUs. The RCS architecture assumes pipelined organization of video-frame acquisition, processing, and result transmitting. The timing diagrams of the aforementioned pipelined processes are depicted in Figure 10.10.

In this implementation the entire video frame—*Frame(i)* is accumulated first in the frame buffer. Then, the process of video-data executions of this frame starts. Since this process is cyclic, the FCR resources in partially

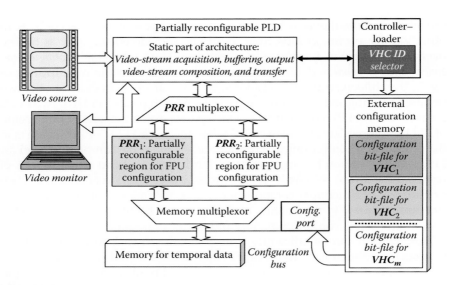

FIGURE 10.9
The RCS with *tandem* organization of cyclically partitioned resources for stream-processing applications.

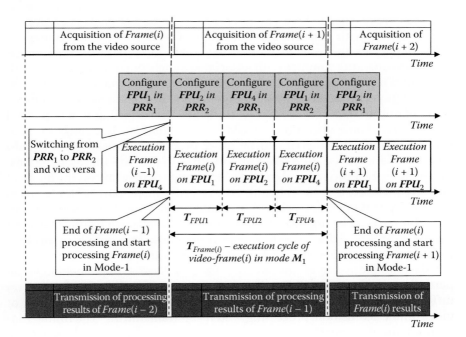

FIGURE 10.10
Timing diagrams of the pipelined processes of data acquisition, data execution, *PRR*s reconfiguration, and result transmission in the RCS with *tandem* on-chip architecture.

reconfigurable region #1 (PRR_1) are already configured to the architecture of FPU_1 after loading the configuration bit-file of the VHC_1 from the external configuration memory to the area of configuration memory at the PLD associated with PRR_1. When the data execution of the *Frame* (*i*) starts on FPU_1, the configuration of PRR_2 for FPU_2 starts simultaneously by loading the configuration bit-file of VHC_2 to the configuration memory area associated with PRR_2. While reconfiguration of the PRR_2 to FPU_2, the results of data-processing of *Frame* (*i*) on FPU_1 are stored in memory for temporal data. When execution of data on FPU_1 is completed as well as configuration of the FPU_2 in PRR_2 of the PLD, the control circuits are switching the FPU multiplexor and memory multiplexor from PRR_1 to PRR_2. This operation usually takes not more than 1–2 clock cycles and thus is negligible for the data-execution process. Now the data-execution process continues by reading the data from the temporal data memory to the FPU_2 deployed in the PRR_2 of the PLD. The results of this execution process are still stored in the temporal data memory. To avoid bus hazards on the temporal data memory, this memory usually is implemented as dual-ported memory. This organization allows reading source data from the memory and writing results to the memory simultaneously (in the same clock cycle). Meanwhile, the reconfiguration of the PRR_1 is performed for FPU_4 as required by model of system operation. When data execution on FPU_2 is completed, FPU_4 should be ready to continue the data-processing. The control circuits allocated in the fixed architecture part are switching the PRR multiplexor and memory multiplexor from PRR_2 back to PRR_1. And then, the data-execution process goes on to FPU_4 configured in PRR_1 and so on. When the data execution for *Frame* (*i*) is completed, the transmission of result starts to the video-output circuits. At this time, *Frame* (*i* + 1) is ready in the buffer and the process continues until the end of the video-stream acquisition.

As it is possible to see in Figure 10.10, there is no reconfiguration timing overhead. The processes of data acquisition, data-processing, and result outputting are going nonstopped and without any stall periods. However, the 50% hardware overhead still exists because one of the identical PRRs is always under reconfiguration and thus excluded from data-processing. Nevertheless, this architecture allows reaching the highest performance for this class of RCS.

For example, instead of just 720 p 60 fps considered in the previous example, the aforementioned *tandem* architecture can achieve stereo-vision (3D) acquisition and processing of XGA video frames on the same 60 fps frame rate with on-chip clock frequency equal to 285 MHz. Indeed, the XGA resolution is 1024 × 768 pixels × 32 bits per pixel and the clock period is equal to 3.5 ns. In case of fully pipelined data paths in each of the FPUs, the complete video-frame data will be executed in 1024 × 768 × 3.5 ns = 2.752 ms. If this time is enough to configure FPUs in PRRs of the PLD, the total time for frame processing will not exceed 3 × 2.752 ms = 8.257 ms. This timing enables reaching 120 fps for mono XGA video stream or 60 fps for stereo-XGA video stream in case video frames from right and left video sensors are processed sequentially (one-after another). The RCS utilizing the method of temporally integrated function-specific components using time-multiplexed PLD resources can be called dynamically reconfigurable macro-processors because the VHCs represent FPUs with component granularity of relatively complex macro-functions.

10.7 ASVP with Spatially Integrated Hardware Components

In some cases, there could be very strict requirements for processing the data rate for multimodal or multitask applications combined with a relatively short mode-switching time. As well, application may require that no one data-stream execution process can be stalled or interrupted during mode switching period. The minimization of hardware cost or power consumption combined with high system reliability can be additional but mandatory specification requirements.

The highest performance can be reached in RCS with statically integrated components presented in Section 10.4. Unfortunately, this class of RCS cannot provide rapid mode switching due to necessity for complete reconfiguration of PLD. Also, complete reconfiguration does not allow seamless mode switching because usually complete reconfiguration stalls all ongoing (in PLD) processes including data acquisition and result outputting.

The alternative solutions are associated with dynamic or run-time integration of components to the processing system. The ASIP approach or RCS based on time-multiplexed PLD resources presented in Sections 10.5 and 10.6, respectively, can provide relatively short mode-switching time and seamless mode-switching process. Unfortunately, the ASIP architecture can provide the highest data-execution rate only for limited number of task segments. On the other hand, the RCS with temporarily or cyclically integrated hardware components have limited speedup due to nature of time-multiplexed resources for sequentially activated FPUs. Another problem with this class of RCS may be relatively low resource utilization factor in the FCR. The hardware and timing overheads discussed in Section 10.6 eventually affect not only productivity but also the cost-efficiency of these RCS. The *tandem* organization of RCS minimizes (or even eliminates) timing overhead and thus accelerates system performance especially for stream processing applications. Nonetheless, the hardware overhead could be even higher because at least 50% of dynamically reconfigurable resources always are in process of reconfiguration for the next (in cycle) FPU.

Therefore, increasing cost-efficiency of the RCS with dynamically reconfigurable resources dictates associated increase of number of PRRs actively used by multiple FPUs working simultaneously when only one or few PRRs are in reconfiguration process.

For example, if the dynamically reconfigurable resources in the FCR consist of 10 separate PRRs when only one of those PRRs is on the reconfiguration process, the resource utilization can reach 90% of these dynamically reconfigurable portion of resources. This can be compared to *tandem* organization of the FCR in partially reconfigurable PLD where only 50% of reconfigurable resources can be used at any time of task execution.

Furthermore, multiple PRRs used for multiple FPUs simultaneously processing data enable multitasking or running multiple SoCs in the FCR. In other words, multiple ASVPs can be configured in temporal and special domains at the same on-chip level of the FCR.

The previously mentioned reasons dictated further development of distributed parallel architectures at the on-chip level of the RCS associated with spatially integrated hardware components in addition to temporal integration discussed in the previous section.

Indeed, the only way for further acceleration of the data-execution process is increasing the spatial parallelism in processing data paths. Therefore, in addition to temporally integrated hardware components, the integration of hardware components in spatial domain should be considered.

In other words, instead of one or two PRRs reserved for FPUs the FCR can be divided on many PRRs for different FPUs and associated communication infrastructure between those PRRs. These on-chip PRRs can be identical in area and the set of resources included in such PRRs. Each PRR is reserved for one FPU encoded in respective VHC configuration bit-file [14]. Similar concept of identical functional units to be configured in the identical PRRs was utilized in the Erlangen slot machine [15]. These PRRs were called "slots." This approach enables constructing the complex computing data paths from identical *bricks* at the on-chip level of the FCR. In turn, it allows simplification and acceleration of SoC integration. The difference between system levels of spatially integrated (plugged in) hardware modules is that the on-chip hardware modules (virtual components) can be stored in virtual form (configuration bit-files) when the pluggable modules at the system level are physical devices. The material costs of those modules as well as the costs of storage, placement, and integration are incomparably higher for physical modules than their virtual analogs. Due to this fact, number of functional components stored in system in virtual form of configuration bit-files can be in orders of magnitudes higher than number of modules presented in physical form. Therefore, flexibility, adaptability, and reliability of the computing system utilizing the VHCs are much higher than the same characteristics of systems utilizing the physically implemented hardware components in the form of functionally specific modules. Furthermore, different variants of the same type of VHC can represent FPUs in different forms of functionality providing different levels of performance depending on available area, power consumption, reliability factors, etc. In turn, this allows rapid adaptation of the RCS to the variations in workload and environmental constraints. Also, the RCS with spatially integrated hardware components can provide the ability for self-restoration from transient and permanent hardware faults by run-time relocation of VHCs from the damaged on-chip slot to the reserved one. This procedure can be performed automatically and in run-time. In contrast to this, the plug-and-play technology at the system level requires manual replacement of physical modules in on-board slots. Also, the manual replacement of hardware modules takes much longer time than at the on-chip level and causes task interrupts or even system restart.

Let us consider the organization of the RCS architecture that can provide the aforementioned features.

In general, the platform for the on-chip implementation of the ASVP with spatially integrated hardware components is the PLD that can provide run-time partial reconfiguration of its PRRs. The FCR of this class of PLDs must accommodate three parts:

1. *The static part of architecture* similar to architecture organization of the RCS with temporal partitioning of resources on function-specific FPUs discussed in Section 10.6. This part of architecture is responsible for interfacing to external I/O devices, on-chip clock generation and distribution, memory subsystem, resource management, and other functions that should stay the same during the period of the task execution.
2. The on-chip communication infrastructure that provides the required bandwidth for the on-chip information exchange between FPUs deployed in the slots of the FCR.
3. The dynamically reconfigurable part of architecture organized as a set of PRR-based on-chip slots. These slots allow dynamic accommodation of the required set of FPUs. Each FPU can be configured in the addressed slot by loading the configuration bit-file of the VHC associated with the FPU to the configuration memory frame(s) of this slot.

In addition to the previously mentioned parts, the resource management subsystem is the mandatory part of this system. The resource management assumes dynamic place and route of FPUs in available slots as well as relocation of FPUs in case of hardware faults or mode changes. This subsystem can be allocated at the on-chip level of the FCR or at the on-board/system level external to target PLD.

The generic organization of RCS with run-time, spatially integrated hardware components (FPUs) is depicted in Figure 10.11. As of this organization any required architecture consisting of multiple FPUs could be allocated in slots according to the specific application requirements.

At start-up time, the initial SoC must be loaded to the PLD. This initial SoC architecture must include the static part of the SoC, interslot communication infrastructure, and blank slots. The blank slots are predetermined PRRs loaded with *dummy* components.

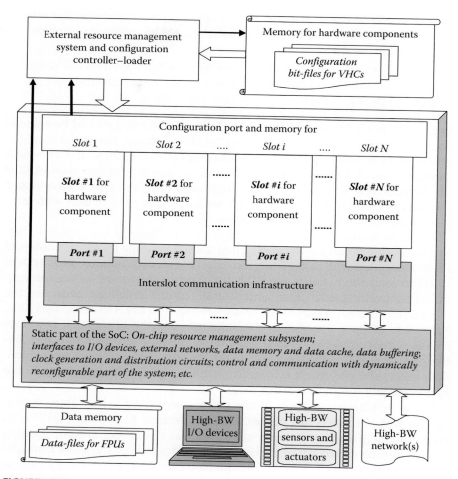

FIGURE 10.11
The generic organization of RCS allowing spatial integration of FPUs in the FCR.

The initial configuration creates the on-chip application-specific infrastructure and reflects the following:

1. Specifics of associated input/output devices and their performance characteristics: set of sensors and actuators and their communication bandwidth, data-exchange protocols, etc.

2. Organization of memory hierarchy: a on-chip buffering and data-cache mechanisms, data-exchange protocols with external memory units, etc.

3. Granularity of slots according to the granularity of the function-specific hardware components (FPUs). This granularity directly

depends on the implementation of application-specific functions. In general, there could be slots with different dimensions and communication bandwidths.

4. On-chip communication infrastructure for interslot data exchange. This infrastructure may have different organization from simple on-chip buses to complex-networks on chip (NoCs).

In general, ports for each FPU-slot (implemented in PRR) should be embedded parts of the aforementioned infrastructure. Hence, the initial configuration of FCR accommodating this infrastructure requires the complete configuration of the target PLD(s). Since PLDs selected for this class of RCS can be relatively large, the initial configuration process may take a longer time in comparison with the relatively smaller PLDs used in RCS with temporal integration of hardware components.

When the initial configuration of resources in the FCR is completed, further accommodation for the application should be done by configuration of PRRs associated with slots. Thus, the external resource management system starts to send the requests for FPUs involved in the task(s) to the configuration controller–loader. The configuration controller–loader generates the respective VHC addresses to configure the required FPUs in associated slots. When this stage is completed, the links between the FPUs must be established. After completion of both previously mentioned operations, the control unit deployed in the static part of the SoC synchronizes data acquisition, data execution, and result storage/transmission processes according to the synchronization signals (Figure 10.12).

Let us consider an example of this process according to the same application as was described in Chapter 9 and in Section 10.4 where ASVP with statically integrated hardware components was discussed. The partitioning for task segments according to the task functions is kept the same throughout all examples in Chapters 9 and 10. However, in the considered case as well as in the case discussed in Section 10.6, each task segment (function) is implemented in the form of a hardware component. Each component being encoded in the configuration bit-file as the VHC is stored in external memory for VHCs. And each VHC_i, $i = 1, 2,...,m$ is designed to fit the appropriate slot (or type of slots). Thus, in case, Mode 1 is activated, VHC_1, VHC_2, and VHC_4 must be loaded to the available slots. Let us assume that $Slot$ 1 has been assigned for VHC_1, $Slot$ 2 for VHC_2, and $Slot$ 3 for VHC_4 as depicted in Figure 10.13.

After loading the configuration bit-files of mode-specific VHCs to the frames associated with assigned slots, FPU_1 will be configured in the area of $Slot$ 1, FPU_2 in the area of $Slot$ 2, and FPU_4 in the area of $Slot$ 3. Obviously, this configuration process will require much shorter time than complete PLD reconfiguration due to smaller volumes of configuration bit-files of loaded VHCs. Then, a resource management subsystem can establish mode-specific interconnects (links) between FPUs according to the predetermined list

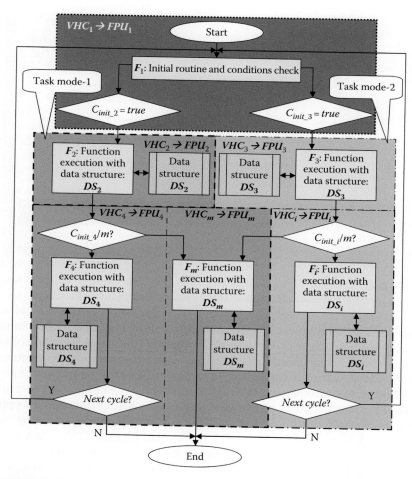

FIGURE 10.12
Partitioning the task on VHCs according to functions.

of interconnects stored in on-chip or external memory. As was mentioned earlier, the inter-slot communication infrastructure can be implemented in many different forms. One of the forms of inter-slot communication infrastructure can be on-chip peripheral bus. Each slot can accommodate the IP-core of the appropriate FPU. The methodology of such design is well known and often used in embedded SoCs (e.g., Xilinx EDK [16]). In case of bus, the links between slots are established in temporal domain where at any period of time only one transaction between two nodes (ports of the slots) is possible. This, however, is the main drawback of this type of inter-slot communication because it can become the performance bottleneck. To overcome the problem of limited bandwidth, different (NoC) topologies can be considered [17]. In case of NoC, multiple transactions between the nodes are

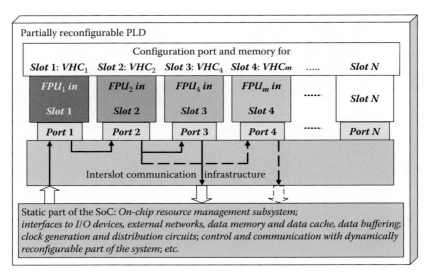

FIGURE 10.13
Example of VHCs distribution in the slots of the partially reconfigurable PLD.

possible to do simultaneously. Determination of the type of interslot communication infrastructure is one of the key points of design for this class of RCS.

For example, considered earlier, let us assume that the on-chip communication infrastructure is implemented as a NoC based on a crossbar switch. This organization allows establishing point-to-point links between each individual FPU slot. In case of pipelined data path, the source data being received from the sensors can be sent over the input circuits deployed in the static part of the SoC to *Port* 1 where FPU_1 is allocated (Figure 10.13). Then, results are transferred to FPU_2 over the link between *Port* 1 and *Port* 2. The link between *Port* 2 and *Port* 3 allows direct transfer of FPU_2 results to FPU_4 located in *Slot* 3. Finally, the results of data execution on the data path composed of by FPU_1, FPU_2, and FPU_4 are transmitted to the output(s) of the SoC.

Let us analyze the mode-switching time on the considered example. Conceptually, there can be two scenarios in accommodating the SoC to the new mode: (1) switching as a reaction on request and (2) switching with mode prediction. Obviously, switching as reaction on request is simpler in implementation and should be considered in case when the next mode is unpredictable. In this case SoC must wait, while all FPUs required for the new mode are configured in selected (by resource management subsystem) slots. After that, the new set of links should be established and activated. This waiting period directly depends on the number of new FPUs to be included in the data path required on the new mode. This period also depends on volume of associated configuration bit-files of requested FPUs. In contrast to that, link switching process requires almost negligible time in comparison to FPU reconfiguration period. Thus, reconfiguration time usually brings the main timing overhead preventing in many cases seamless (to application) mode switching.

The ability to predict the upcoming mode allows significant reduction of mode-switching time because the configuration of FPUs in spare slots (for most probable mode) can be done, while the current mode is active. Thus, when request for the predicted mode arrives, it is necessary to switch the links between slots to accommodate the data path for the new mode. Both of the aforementioned cases are illustrated in Figure 10.13. As discussed earlier, in case of Mode 1, FPU_1, FPU_2, and FPU_4 are configured in *Slot* 1, *Slot* 2, and *Slot* 3, respectively. However, if the probability for switching to function— F_m according to condition C_{init_4}/m is high, the new mode requiring FPU_1, FPU_2, and FPU_m most probably can be requested. Therefore, if there is a spare slot (e.g., *Slot* 4) in FCR, the VHC_m can be loaded in the configuration memory associated with this slot to configure the FPU_m in *Slot* 4 prior to the possible mode request. In other words, configuration memory segment associated with spare slot(s) plays the role of cache memory for VHC(s). If the C_{init_4}/m selects function F_m instead of F_4, the only what is necessary to do is to change the topology of links between slots. The set of links between components for Mode-1 is shown in Figure 10.13 as solid lines composing a pipelined data path based on FPU_1, FPU_2, and FPU_4. When new mode requires a modified data path composed of FPU_1, FPU_2, and FPU_m, the set of links must also be modified to new set shown in Figure 10.13 as dotted lines. The changes in point-to-point links between ports associated with respective slots can be done in many ways. Some FPGA devices with column-based microarchitecture (e.g., Xilinx Virtex-E or Virtex-2/2 Pro) included special tri-state buffers for switching the aforementioned interconnects. The recent tile-based FPGA microarchitectures allow utilization of the on-chip multiplexors for this purpose (e.g., Xilinx Virtex-4/5/6 and 7). The multiplexor-based crossbar communication infrastructure interconnecting the initially configured static part of the SoC and four dynamically reconfigurable slots can be considered as an illustration to the earlier example. This organization is shown in Figure 10.14. Each port consists of one multiplexor that can connect the input bus of this port to any output bus of other ports in the system or to the ground (all input bits are equal to zero). When the multiplexor input bus is connected to the ground, the associated slot is assumed disconnected from the rest of the system.

The output data bus of each port is associated with the respective input data buses of multiplexors: output bus of the Port1 to input buses #1 of all multiplexors, output bus of Port 2 with input buses #2 of all multiplexors, etc. Each port is dedicated to the respective slot or static part of the SoC. Each input multiplexor has its port-selector circuits connected to the resource management and mode-switching subsystem(s). For example, considered in Figure 10.13, the implementation of Mode 1 on the aforementioned communication infrastructure will look as follows: (1) all slots are configured similar to Figure 10.13 and (2) the data links required for the pipelined data path

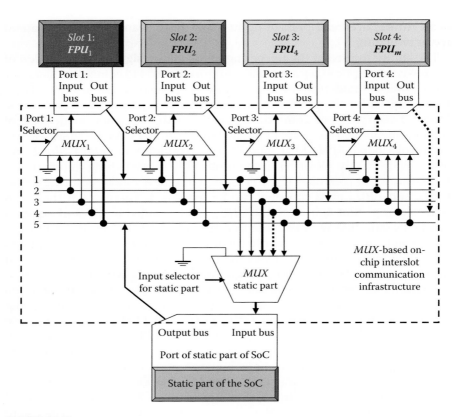

FIGURE 10.14
Crossbar organization of communication infrastructure based on multiplexors.

according to the Mode 1 are depicted by bold solid lines. Thus, port selectors
are set to switch:

1. MUX_1 to bus #5 associated with the output bus of the static part of the
   SoC and thus bringing input data for execution on FPU_1 deployed in
   Slot 1.
2. MUX_2 to bus #1 transferring the results from FPU_1 to the FPU_2
   inputs.
3. MUX_3 to bus #2 allowing FPU_4 located in **Slot 3** to receive results of
   upstream processing on FPU_2.
4. MUX of static part to bus #3 that connects the static part inputs to the
   outputs of FPU_4 and receive final results of the pipelined data path.

If the next mode can be predicted with high probability, the requested FPU(s)
could be deployed in the spare slot(s). In an example shown in Figure 10.13,
FPU_m was configured in **Slot 4** ahead of time according to the probability

of switching the functions from F_4 to F_m (as per Figure 10.12). The process of configuration of FPU_m in the selected *Slot* 4 requires prevention of structural hazards (conflicts in hardware circuits) in the communication infrastructure.

To exclude any influence on input circuits, the associated multiplexor—MUX_4 should be set to the grounded inputs and thus, no FPU can influence on the inputs of *Slot* 4. At the same time, no active FPUs is connected to bus #4 associated with the outputs of *Slot* 4. When configuration of FPU_m is completed by loading the configuration bit-file of the VHC_m to the configuration memory associated with *Slot* 4, the system is ready to switch to function F_m instead of F_4 when necessary. The request of mode switching initiates the port-selector circuits for appropriate changing of the set of links between the slots. This process, however, cannot start before completion of the last cycle of data execution in Mode 1. When the control circuits have detected this event, all multiplexors can be switched to the grounded inputs to prevent any incorrect (noise) data on port inputs. After that the resource management subsystem sends appropriate signals to the port-selectors to switch MUX_1 to bus #5, MUX_2 to bus #1, and MUX_3 to ground (disconnecting FPU_4 from the data path). Then, MUX_4 should be connected to bus #2 creating the link between FPU_2 and FPU_m (shown in by the dotted line in Figure 10.14). Finally, the output of *Slot* 4 over the bus #4 should be connected to the output buffers in the static part of the SoC via *MUX* static part as by the dotted line in Figure 10.14 too.

It is necessary to mention that, switching of multiplexors to the initial (grounded) position and then to positions required by the new mode of operation takes a relatively short time (usually in the range of hundreds of nanoseconds). This time is similar in order of magnitude to the mode switching time in the ASIC-like (one-time configurable) circuits deployed on PLDs (discussed in Section 10.4). Also, the performance of the considered class of ASVP with spatially integrated hardware components is also the same as the performance of ASIC-like SoCs due to the implementation of the fully pipelined data path according to the same mode of task operation.

Hence, ASVP with spatially integrated hardware components can provide the same performance and mode-switching timing in case the next mode could be predicted consuming much less hardware resources for its FCR.

Let us estimate the performance of this SoC on the example considered earlier using the same video-processing task as was discussed in Section 10.6. In this example, video-frame organization (1280 × 720 pixels) was considered as the data structure for all functions involved in task execution. The on-chip clock frequency was determined for 333 MHz (3 ns/clock cycle). Assuming that a fully pipelined data path consists of FPU_1, FPU_2, and FPU_4 or FPU_m, the cycle time can be reduced up to one clock cycle. According to these assumptions, each video frame can be processed in 1280 × 720 × 3 ns = 2.765 ms which is equal to the frame rate of 360 frames per second.

It is necessary to mention that if the data path of each FPU is also fully pipelined and aligned to the cycle time equal to one clock cycle, the frame rate will not depend on the number of FPUs in the pipeline. This is because combined data paths can be considered as one long data path consisting of all processing stages included in FPU_1, FPU_2, and FPU_4 or FPU_m. Thus, the number of functions in this combined data path influences only the latency (from a performance point of view) and the number of slots involved in the data path (from an area/cost point of view). From a application point of view, the extended latency should not be a critical aspect for consideration. Indeed, even in case of hundreds of stages in the data path composed of a set of FPUs, the latency will be in the range of units of microseconds. This can be compared to the vertical synchronization period (5 lines between video frames) equal to 115 µsec or even to the horizontal synchronization period (80 pixels) equal to 1.44 µsec according to the 720 p 60 fps standard. These periods of time are much longer than latency and/or the previously-mentioned mode-switching time required for multiplexors.

As was discussed in Section 10.3, integration process of FPUs with static part of the ASVP requires proper synchronization. The organization of this synchronization in ASVP with spatially integrated hardware components is the most complex process in comparison to previously discussed methods of integration because it consists of three inter-dependable levels of synchronization:

1. *Synchronization of slot reconfiguration process.* On this level of synchronization, the control and resource management subsystems are conducted:
   a. Selection of VHC and target slots for reconfiguration.
   b. Initiation/termination of VHC configuration process according to the up-coming mode of operation.
   c. Updating resource utilization information: used and spare slots, placement map, etc. This level of synchronization insures placement of all FPUs in appropriate moments of time and within available periods of time.

2. *Synchronization of the mode-switching process.* At this level of synchronization, the following control and resource management subsystems are conducted:
   a. Determination of the moment for initiation the mode-switching process according to the completion of execution of the current mode
   b. Readiness of the resources in the target slots for the mode switching
   c. Readiness of the communication infrastructure
      Thus, this level of synchronization provides proper timing for runtime assembling of all components into the ASVP data path optimized for the upcoming mode of operation.

3. *Synchronization of the data-execution process.* Determining the moment(s) of time when each component of the ASVP starts its operation and completes it according to the assigned data structure specifics and execution process timing of the upstream and downstream components in the data path

It is assumed that data structures associated with an application can provide boundary indicators for their formats. The examples could be (1) for video-processing applications: vertical (frame valid) or horizontal synchronization signals, and (2) for communication tasks: start/stop signals in serial communication protocols, packet synchronization signals/synchronization bytes in transport stream packets, etc. These signals can be used by the control circuits in the static part of the ASVP for initiation of data execution processes in the components. In case of multicomponent pipelined data-path organization, the inter-component handshaking signals (flags) also could be used for data-flow synchronization during the data-execution process.

All the previously mentioned aspects of spatial integration of components in the ASVP during the mode preparation, initiation, data execution, and termination processes should be considered in the design of the resource management and/or control subsystems in this class of RCS. The organization of the resource management and resource scheduling subsystems is the wide area of research for the class of multi-FPU and multicore RCS.

There are three approaches for the implementation of the aforementioned resource reconfiguration, scheduling, and synchronization mechanisms:

1. *Centralized mechanism based on the CPU core*: In this approach, the CPU-based core performs all control, communication, and synchronization functions dealing with

    a. Controller–loader during the slot configuration with appropriate VHCs

    b. Communication infrastructure to set up the new set of links and create the required data-path architecture

    c. Synchronization circuits for proper scheduling the data-execution initiation, termination, and data-transaction processes
    This method allows relatively simple implementation of all previously listed functions and procedures but can cause a problem in case the mode-switching time is a critical parameter due to the limited performance of microprocessing cores.

2. *Centralized mechanism based on dedicated circuits*: In this approach, all the aforementioned functions that need to be accelerated according to the short mode-switching time requirements are implemented by special dedicated circuits designed just for rapid reconfiguration of the slots, links reconfiguration, and run-time synchronization. This approach can bring significant additional time for design

and implementation of these mechanisms but allows reaching the shortest time in mode switching and adaptation to the workload variations.

3. *Decentralized mechanism based on dedicated circuits included in each FPU:* In this approach, each FPU should contain circuits responsible for run-time assembling, scheduling the resources. Thus, levels of synchronization are distributed in FPUs, ports of slots, and communication infrastructure that may have special buses to exchange control and synchronization information between components allocated in slots. In other words, this method allows self-assembling of the mode-specific data path of the ASVP by collaboration between components, each of which contains dedicated circuits for this purpose. This is the most complex way of implementing the ASVP with spatially integrated hardware components. However, for cases where utilization of centralized approach is limited due to requirements for high dependability and ability for self-restoration, this method could be almost only possible solution.

In general, implementation of the resource management, resource scheduling, and synchronization subsystems requires on-chip level (in static part of the ASVP) and system level stand-alone devices (based on microcontroller(s), CPLD(s), or FPGA devices). Certainly, there could be many variations of the architectures of these subsystems according to the application specifics of the respective RCS. Therefore, it is almost impossible to observe all methods and ways for utilization and implementation of the ASVP with spatially integrated hardware components. Furthermore, this is one of the widest areas of research of dynamically partially reconfigurable computing systems.

10.8 Summary

In this chapter, one of the most important aspects of architecture integration in the FCR of the RCS has been considered in detail. This aspect is unique for the RCS because in computers with fixed hardware architecture, integration of all architecture components is done on the design stage and then never required again. In contrast to that, the concept of architecture reconfiguration as the mechanism for architecture-to-workload adaptation requires run-time variations of components functionalities and topology of the links between components. Therefore, the concept of component virtualization discussed in Chapter 9 logically was expanded to the concept of virtualization of the entire computing architecture. In other words, the architecture virtualization looks like the higher level of component virtualization when the entire ASP is not present in hardware circuits at all times. Instead, only a subset of components is placed in the FCR according to the functional requirements of the task mode. Respectively, the set of links between components associated with this mode should be established in the FCR prior to the mode-switching time. Therefore, the problem of synchronization of processes associated with architecture reconfiguration became an important consideration. These synchronization processes are associated with (1) component placement and reconfiguration, (2) establishing an appropriate set of interconnections between components, (3) initiation and termination of data-execution processes, and (4) synchronization of data flow in the multifunctional data path.

All these aspects have been analyzed in this chapter in light of the organization of the application-specific virtual processor (ASVP) composed by the integration of function-specific processing units (FPUs). The concept of ASVP assumes that only those FPUs that are required for the current task mode are placed in the FCR. This placement is conducted by loading the configuration bit-files of associated VHCs to the determined areas of configuration memory of the target PLD(s). Then, the configured FPUs can be integrated with the static part of the ASVP creating the mode-specific processing architecture.

There could be different ways of the aforementioned integration. The simplest way assumes off-line (static) integration of components on the design stage. In this case, ASVP will consist of several mode-specific processors. Each processor being stored in the form of a configuration bit-file can be initiated according to the respective mode by loading (completely) to the configuration memory of the target PLD. In the case of a new mode, the associated mode-specific processor will be configured instead of the previous one. This approach is effective in cases when the number of modes is relatively small and the available mode-switching time is greater than reconfiguration time of the target PLD(s). Alternatively, the run-time or dynamic integration is required. In this case, resources of PLD(s) can be partitioned on the required FPUs in the temporal or spatial domain. Both classes of RCS have been discussed showing the pros and cons of these methods of integration.

The approach known as ASIP assumes utilization of a CPU-based instruction processor as the core of the ASVP. The virtual software components (VSCs) as software program modules are executed on this CPU-core executing algorithmically intensive segments of the task. However, computationally intensive task segments are executed on hardware accelerators dedicated function-specific processing circuits (FPUs). In this regard, certain reconfigurable resources are reserved for the associated FPU(s). The integration aspect here looks like a plug-and-play mechanism when the requested function-specific hardware circuit can be configured in the target area of the PLD and interfaced with the instruction processor over a dedicated port and bus. This approach is very effective when a majority of task segments are algorithmically intensive, but there are some task segments that require much higher computation performance.

When most of task segments are computationally intensive and execution of data on the instruction-based processors cannot satisfy performance requirements, only FPU-based data paths can be used in the ASVP architecture. Two possible methods can be utilized in the composition of these RCS architectures: (1) methods based on the sequential (temporal) configuration of FPUs in the available FCR and (2) methods based on the spatial partitioning of resources on multiple FPUs creating parallel mode-specific architecture.

The first of these methods is more economical in comparison with alternative methods associated with spatial partitioning of reconfigurable resources. Nevertheless, these methods allowed executing almost any computationally intensive tasks on relatively small PLDs providing performance higher than the instruction processor. Integration of FPUs with each other in these methods had to be done over the temporal data memory. Thus, the integration was insured by identical FPU-to-memory interface included in the design of any FPU used in this class of RCS. Certainly, RCS utilizing the aforementioned method of resource partitioning could provide lower performance characteristics than RCS with parallel processing architectures.

The RCS with spatial resource partitioning can reach the highest performance comparable with ASIC-like systems requiring much less resources than the same systems with *monolithic* designs. However, the dynamic nature of component integration in their architectures is the most complex in comparison with all other classes of RCS. Thus, one of the most complicated parts of these systems is resource management and configuration control subsystems responsible for all levels of ASVP integration and synchronization of reconfiguration and data-execution processes. Therefore, this class of RCS still is in the process of research and development but certainly has the highest potential in practical applications in many areas of computing technologies.

Exercises and Problems

Exercises

1. What are the two main conceptually different parts of the RCS architecture in general form from a data-processing point of view?
2. What types of components compose the data-processing part of the RCS architecture in general form?
3. In which form can sequential instruction processors be implemented in the RCS architecture?
4. In which part of the data-processing system can VHCs be implemented?
5. What is the main difference between I/O interfaces associated with (a) instruction-based microprocessor cores and (b) field of configurable logic resources?

6. Describe the three main parts of memory hierarchy in the general RCS architecture?

7. Why must the cache memory for software components be separated from cache storing the VHC configuration bit-files?

8. Define the application specific processor (ASP).

9. In which possible forms can the ASP be implemented? What are the reasons for implementation of the ASP in each of the forms?

10. Why are some function-specific processing units (FPUs) static in ASP and why can there be dynamic FPUs?

11. Define the application-specific virtual processor (ASVP). Why are there static and dynamic parts in the ASVP architecture?

12. What are the two processes that may go in parallel or one after another in hardware resources allocated for an ASVP?

13. Describe the two main methods for the integration of function-specific processing units (FPUs) into mode-specific ASP. What is the main difference between these two methods? What are the benefits and main drawbacks of these approaches?

14. In case of static integration of components into mode-specific ASP, what stages are needed and why?

15. When is the static integration method of the ASP architecture effective? List and describe the two main conditions.

16. What is the main difference of runtime (dynamic) integration of the ASP architecture in comparison to the static integration process?

17. When is the dynamic integration of ASPs in FCR needed and most effective?

18. What is the main drawback of the run-time ASP integration method in comparison to the static integration of the ASP architecture?

19. What are the main stages of on-line run-time integration of FPUs into the complete ASP architecture in the FCR?

20. Describe the three main methods for the run-time integration of the ASP architecture in the FCR of the RCS.

21. How can the multimodal ASVP be implemented in the statically reconfigurable RCS? Which part(s) of the application should be implemented as a static component in the hardware circuit and which segments of application should be implemented as VHCs?

22. Define the class of application-specific instruction set processors (ASIPs). Why does the CPU-based computing system need hardware accelerator(s)?

23. In which cases is the ASIP with reconfigurable hardware accelerators a more efficient solution than ASIP with statically deployed hardware accelerators?

24. List the timing overheads that may decrease the efficiency of the ASIP with reconfigurable hardware accelerators. What is the condition when these overheads can be considered as negligible?

25. Define the concept of application-specific virtual processor with temporarily integrated hardware components. In which cases can utilization of this concept in the RCS be efficient?
26. Which components should be included in the static part of the ASVP with temporarily integrated hardware components and which components should be kept in the dynamic part of this RCS?
27. What is the main benefit of the RCS architecture employing temporal partitioning of the hardware resources on the task segments? What are other important benefits of this class of RCS architecture?
28. What is the main drawback of the RCS architecture with temporal partitioning of resources on task segments?
29. Is there any architectural solution to minimize or eliminate the timing overhead associated with the reconfiguration time in the RCS with temporarily integrated components?
30. What is the hardware overhead in the FCR in case the *tandem* architecture is used for the RCS with temporarily integrated components?
31. In which cases can the ASVP with spatially integrated hardware components be considered as the most effective architectural solution? List the most important conditions.
32. What are the benefits of the ASVP with spatially integrated hardware components? Is it possible to perform parallel multitask execution on this class of RCS?
33. Why segmentation of the on-chip FCR on identical PRRs allows simplification and thus, acceleration of integration process of the SoC?
34. What are the differences between SoC integration using VHC-based FPUs and the integration process on system level based on plug-and-play modules? What are the benefits and drawbacks of both approaches?
35. List the three main parts of the architecture that the FCR of the RCS must accommodate at all times in case of implementation of the ASVP with spatially integrated hardware components.
36. Why does the implementation of the ASVP with spatially integrated hardware components requires initial configuration? And what parts of the ASVP architecture must be included in this initial architecture configuration?
37. What is the reason why the ASVP with spatially integrated hardware components can provide performance much higher than ASVP with temporarily integrated components and close or equal to performance of ASIC-like PLD-based systems?
38. When (in what conditions) the mode-switching time in the ASVP with spatially integrated hardware components can be the same as in ASIC-like PLD-based systems?
39. Why is the ASVP with spatially integrated hardware components a more economical solution than statically integrated RCS or ASIC-like systems?

40. List all levels of synchronization required for proper and conflict-free integration of FPUs in the data-path of the ASVP with spatially integrated hardware components. Describe the functions associated with each of level of synchronization.
41. List the three main approaches for the implementation of resource reconfiguration, scheduling, and synchronization mechanisms. Describe the pros and cons of each of them.

Project

The embedded computing system has to perform the multimodal video-processing application.

The specification of the algorithm segmentation and data structure is as follows:

1. Segmentation diagram of the workload is shown in Figure P1.
2. The data structure of the video frame is according to 1080 p × 60 fps standard (frame rate is equal to 60 video frames per second progressive scan).
3. System can perform video processing in one of the *eight* possible modes.
4. Mode-switching time should not exceed one video-frame period.

Determine the best way of system implementation after the analysis of

1. Implementation in one statically configured FPGA device, which accommodates logic for all 10 segments: *S0–S9* (Large FPGA ASIC-like SoC solution).
2. Implementation in one FPGA device to be reconfigured to the requested mode according to the event (multimodal statically reconfigurable RCS solution).
3. Implementation in ASIC accommodating all 10 segments: *S0–S9* (ASIC SoC solution);

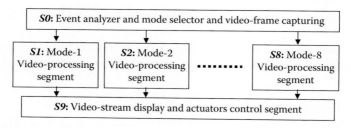

FIGURE P1
Block diagram of segmented algorithm structure.

TABLE P1

Xilinx Virtex-5 LX Family FPGA Product Selection Matrix

FPGA Device	Number of Slices	Configuration Bit-File (bit)	Device Cost in Quantity	
			1,000	10,000
XC5VLX30	4,800	8,374,016	380	150
XC5VLX50	7,200	12,556,672	720	320
XC5VLX85	12,960	21,845,632	1,125	630
XC5VLX110	17,280	29,124,608	1,816	985
XC5VLX155	24,320	41,048,064	3,180	1230
XC5VLX220	34,560	53,139,456	4,788	2590
XC5VLX330	51,840	79,704,832	12,915	7540

TABLE P2

Resource Requirements (in Slices) for Each Segment of the Workload

Segment #	S0	S1	S2	S3	S4	S5	S6	S7	S8	S9
Number of logic slices	1238	4220	4380	2136	4020	3082	1682	4334	3210	1074

Note: Each mode requires execution of *S0, S9, and one of the Si*, where $i = 1, 2, \ldots 8$.

Suggested family of FPGA devices is Xilinx Virtex-5LX. Available logic resources (in CLB-slices) and approximate cost of each FPGA device (in quantity: 1,000 and 10,000 units) is given in Table P1.

The results of the compiled VHDL-code for each segment are collected in Table P2.

There could be possible design solutions for System-on-Chip implementation

1. In a form of large statically configurable FPGA incorporating all required modes of operation

2. In a form of multimodal RCS with one FPGA device to be reconfigured to the requested mode

3. In a form of application-specific integrated circuit (ASIC)

To figure out which implementation of the embedded computing system is the best for the required amount of production, perform the following analysis.

Problem 1 *Analysis of hardware and timing resources*

Solution 1 Implementation in large FPGA in the form of ASIC-like "monolithic" SoC

TABLE P3

Resource Utilization Factor in ASIC-like SoC according to the Mode of Operation

Mode #	M1	M2	M3	M4	M5	M6	M7	M8
RUF								

a. Calculate the amount of total logic resources needed to accommodate complete System-on-Chip in Large FPGA (with all 8 modes implemented in one FPGA)—*Rtotal*

b. Determine the most suitable FPGA device (according to Table P1) that can accommodate this SoC

c. Calculate the value of *Resource Utilization Factor* (RUF) for ASIC-like SoC in the selected FPGA device for each of 8 modes of operation: *M1, M2 … M8* and fill Table P3

d. Calculate the start-up time for this system, assuming that the start-up time is equal to the configuration time for the selected large FPGA.

Determine the volume of the configuration file from Table P1 and calculate the start-up time if the configuration of the FPGA device should be conducted via JTAG configuration port.

Note: JTAG configuration port provides 1 bit per clock cycle with maximum 66 MHz clock frequency.

Solution 2 Implementation in the form of multimodal RCS with one FPGA device

a. Calculate the *maximum logic* resources: number of logic slices for the mode that requires the largest area in the FPGA to accommodate *any one of the eight modes*

  Note: Each mode has its own configuration bit-stream. Each configuration bit-stream encodes *S0 and S9 + one of the Si, where i =* 1, 2,… 8

b. Select the best suitable FPGA device for multimodal RCS (from the Table P1) according to maximum logic required that can accommodate any of the eight mode-specific SoCs

c. Calculate the value of *Resource Utilization Factor* (RUF) for multimodal RCS in the selected FPGA device for each of the eight modes of operation: *M1, M2 … M8* and fill Table P4

d. Determine the best suitable reconfiguration scheme to satisfy specification requirements regarding mode-switching time *less than one video-frame period*

TABLE P4

Resource Utilization Factor in Multimodal RCS
According to the Mode of Operation

Mode #	M1	M2	M3	M4	M5	M6	M7	M8
RUF								

For this selection, calculate FPGA reconfiguration time according to available configuration schemes (provided next), which satisfies specification requirement regarding the mode-switching time: *less than one video-frame period* = 16.66 mS for 60 fps.

The following FPGA configuration schemes are available for a given family of FPGAs:

  a. JTAG configuration scheme—1 bit/c.c. with max 66 MHz clock frequency
  b. SelectMAPx8 configuration scheme—8 bits/c.c. with max. 100 MHz clock frequency
  c. SelectMAPx16 configuration scheme—16 bits/c.c. with max 100 MHz clock frequency
  d. SelectMAPx32 configuration scheme—32 bits/c.c. with max 100 MHz clock frequency

Note: Configuration time = configuration bit-file volume/configuration bus bandwidth.

Problem 2 *Analysis of system cost and cost-efficiency*

Estimate and compare the costs of the system for different implementation solutions and select the most appropriate solutions according to the minimum cost associated with production volume.

Note: The cost of the system must take into account the hardware platform cost on system level plus development and design costs divided by production volume.

In other words, **System cost** = Cost of FPGA or ASIC device + Cost of the board and supporting components + SoC Development cost/number of systems to be produced.

Estimated platform development and SoC design costs are given in Table P5.

$$\text{ASIC device cost} = \text{tooling cost/number of devices} + \text{device}$$
$$\times \text{production cost}$$

For the aforementioned ASIC, *estimated tooling cost* = \$1,000,000 and *device production cost* = \$50.

TABLE P5

Estimated Cost of Platform Development and SoC Design

Solution #	SoC Implementation Platform	SoC Design Cost
1	Large FPGA	$120,000
2	Multimodal RCS	$100,000
3	ASIC	$1,500,000

TABLE P6

Cost of System Unit in Case of 1,000 and 10,000 Units Production

Variant of Design Solution	ASIC-like SoC Implemented in Statically Configurable FPGA	Multimodal SoC Implemented in Reconfigurable FPGA	SoC Implemented in ASIC
Cost of system unit in case of 1,000 units production			
Cost of system unit in case of 10,000 units production			

For example: for series of 1000 ASICs, the ASIC device cost = $1,000,000/1000 + $50 = $1,050 per device.

Estimated cost of the printed circuit board (PCB) production, verification, and supporting components is the same for each variant of system implementation and is equal to $500 per system.

Calculate the cost of system unit produced in quantity of 1,000 and 10,000 units for

i. Large FPGA (ASIC-like SoC solution)

ii. Multimodal SoC solution (multimodal RCS) and

iii. SoC implemented in ASIC

Fill Table P6 for the volume accordingly and determine the best variant of implementation for production volume of 1000 systems and 10,000 systems.

References

1. S. Hauck and A. DeHon, *Reconfigurable Computing: The Theory and Practice of FPGA-based Computation*, 1st ed., Morgan Kaufmann, San Francisco, CA, November 2007, 944pp.
2. J. Cardoso and M. Hubner, *Reconfigurable Computing: From FPGAs to Hardware/Software Co-Design*, Springer, New York, November 2014.

3. M Gries and K. Keutzer (Eds.), *Building ASIPs: The Mescal Methodology*, Springer, New York, June 2005, 375pp.
4. Analog Devices Inc., *ADSP-BF60x Blackfin Processor Hardware Reference*, Rev. 0.5, Analog Devices Inc., Norwood, MA, February 2013, Part Number: 82-100113-01, Chapter 1.
5. R. Razdan, B. Grundmann, and M. D. Smith, Dynamically programmable reduced instruction set computer with programmable processor loading on program number field and program number register contents, US Patent: US5696956 A, Publication Date: December 9, 1997.
6. J. Hauser and J. Wawrzynek, GARP: A MIPS processor with a reconfigurable coprocessor, in K. L. Pocek and J. M. Arnold (Eds.), *Proceedings of the IEEE Symposium on FPGAs for Custom Computing Machines*, IEEE Computer Society Press, Los Alamitos, CA, 1997, pp. 12–21.
7. C. Rupp, M. Landguth, T. Garverick, E. Gomersall, H. Holt, J. Arnold, and M. Gokhale, The NAPA adaptive processing architecture, in *Proceedings of the IEEE Symposium on FPGAs for Custom Computing Machines*, NAPA Valley, CA, April 15–17, 1998, pp. 28–37.
8. Xilinx Inc., Zynq-7000 All Programmable SoC Overview, Datasheet, DS190 (v1.8), May 27, 2015, Product Specification.
9. ALTERA Corporation, User-Customizable ARM-based SoCs for Next-Generation Embedded Systems, White Paper: WP-01167-1.1, June 2013.
10. V. Baumgarte, G. Ehlers, F. May, A. Nückel, M. Vorbach, and M. Weinhardt, PACT XPP—A self-reconfigurable data processing architecture, Kluwer Academic Publishers, *The Journal of Supercomputing*, 26(2), 167–184, September 2003.
11. S. Ganesan and R. Vemuri, An integrated temporal partitioning and partial reconfiguration technique for design latency improvement, in *Proceedings of the Design, Automation & Test in Europe (DATE'00)*, Paris, France, March 27–30, 2000, pp. 320–325.
12. F. L. Kastensmidt, L. Carro, and R. Reis, *Fault-Tolerance Techniques for SRAM-based FPGAs*, Springer, New York, 2006.
13. M. Abramovici, M. A. Breuer, and A. D. Friedman, *Digital Systems Testing and Testable Design*, IEEE, John Wiley & Sons, New York, 1990.
14. L. Kirischian, I. Terterian, P. W. Chun, and V. Geurkov, Re-configurable parallel stream processor with self-assembling and self-restorable micro-architecture, in *Proceedings of International Conference on Parallel Computing in Electrical Engineering*, Dresden, Germany, September 7–10, 2004, pp.165–170.
15. M. Majer, J. Teich, A. Ahmadinia, and C. Bobba, The Erlangen slot machine: A dynamically reconfigurable FPGA-based computer, *Journal of VLSI Signal Processing Systems*, 46(2), March 2007, pp. 15–31.
16. Xilinx Inc., *EDK Concepts, Tools and Techniques: A Hands-on Guide to Effective Embedded System Design*, User guide, UG683 (v14.3), October 16, 2012.
17. G. D. Micheli and L. Benini, *Networks on Chips: Technology and Tools*, 1st ed., Morgan Kaufmann Publishers, San Francisco, CA, July 20, 2006.

Index